# The Cities That Vanished

# The Cities That Vanished

**True Stories of America's Lost Cities**

Charles Kensington

Hidden Histories
Media

Staunton, VA 24401



Printed in the United States of America

First Printing, 2025

Published by Hidden Histories Media



Paperback ISBN: 978-1-971137-10-0

# Contents

# Introduction

Dawn breaks slowly over the Mississippi River floodplain. A thin band of light slides across the horizon, turning the grass a muted gold as the last of the night's mist lifts from the ground. On this quiet morning, the air is still, heavy with the scent of wet earth and distant river water. Birds stir in the reeds. A lone deer moves along the tree line. Nothing suggests that beneath the surface lies the memory of a city that once reshaped an entire continent.

A geophysicist stands beside a set of scanning instruments arranged near the base of a gentle rise. To passing drivers on the nearby highway, the mound looks like nothing more than an oddly shaped hill. To the researcher adjusting cables in the early light, it is a place alive with questions. When the machine hums to life, the monitor flickers, then blooms with faint contours, rectangles, voids, alignments, and signatures of long-buried structures. A thousand years collapse into a single image. What appears on the screen is not a hill at all, but a memory of walls, platforms, ceremonial spaces, and homes belonging to a city that once held tens of thousands of people.

This moment, silent and almost private, captures the spark that guides the pages that follow. The modern world is full of places just like this. Mounds resemble rolling fields. Depressions lie masked by forest. Stone foundations crumble beneath vines. Traces disappear beneath floodplains or pavement. They lie in plain sight, waiting for someone to look differently. And when we do, the land begins to speak. The ground remembers.

## The Question Beneath Every Ruin

Across the Americas, cities once thrived with the noise of markets, the rhythm of rituals, the weight of diplomacy, and the urgency of survival. Their influence stretched across vast territories. Their innovations shaped technologies, trade routes, and cosmologies. Their leaders governed through alliances, ceremonies, and carefully orchestrated displays of power. Their farmers mastered environments as diverse as the Amazon basin, Andean plateaus, desert valleys, and temperate floodplains.

Yet many collapsed. Others transformed beyond recognition. Some disappeared so completely that later generations never even knew they existed.

This book asks a simple, unsettling question. How does a city vanish? How does an entire civilization, with its plazas, its leaders, its garments, and its stories, fade from memory until only fragments remain? And how do the echoes of those cities continue to shape our understanding of America's past?

The answers are rarely singular. Civilizations do not fall like dominos toppled by a single force. They bend, strain, adapt, and endure until the combined pressures of environment, politics, crisis, or contact push them beyond their limits. Their endings often linger for generations, unfolding slowly through migration, environmental change, shifting alliances, internal debates, or devastating epidemics.

Understanding how these cities rose and fell does not simply illuminate the past. It deepens our understanding of ourselves.

## Awe in the Silence of Abandoned Worlds

To stand among the remains of an ancient city is to feel the weight of time in a way that bypasses logic and strikes directly at the senses. A plaza once filled with thousands now sits empty, visited only by wind and grass. A stone staircase leads nowhere. A cliffside dwelling overlooks a canyon so quiet that every footstep seems louder than the stories that once filled its walls. A ceremonial mound rises from farmland, its purpose forgotten, its builders unnamed in written memory.

There is awe in these places, not only in their scale but in their persistence.

Every ruin carries a kind of emotional gravity. It invites wonder, admiration, and a faint ache for the unknown. Who lived here? What brought them together? How did they balance belief and survival? What did they fear? What did they celebrate? What did they lose?

And perhaps more urgently, what remains of their story today, and why did it take so long for us to see it?

Modern archaeology has answered some questions while opening many more. LiDAR imaging reveals geometric patterns beneath rainforest canopies. Soil chemistry exposes ancient fields engineered to control floods. Dendrochronology reconstructs drought cycles that strained Pueblo alliances. Cave excavations uncover mass burials that point to political turmoil in the Moche world. Underwater surveys reveal drowned structures near Lake Titicaca that expand our picture of Tiwanaku. Oral traditions carried through generations restore narratives that material evidence alone cannot explain.

Each discovery reminds us that the past is not static. It grows clearer and more complex with every new piece of evidence.

## The Journey Ahead

This book is not a chronological march through time, nor a technical summary of archaeological findings. It is a journey, site by site and culture by culture, through the cities that vanished across the Americas and the forces that shaped their destinies.

You will travel from the fertile floodplain of Cahokia to the volcanic slopes of ancient Quito, from the high Andean cloud forests to the deep valleys of the Mississippian South. You will see Teotihuacan's Avenue of the Dead glowing under dusk light and stand among the broken warrior columns of Tula as thunder rolls over distant hills. You will watch the Amazon canopy peel back to reveal geometric earthworks and follow LiDAR sweeps across the Maya lowlands, where entire road networks appear like ghosts against dense jungle.

As you move through these landscapes, patterns emerge.

Some cities thrived through mastery of environment, engineering, or trade. Others gained prominence through political unity or charismatic leadership. Many endured drought cycles that pushed societies toward reinvention or conflict. Some fragmented under the pressure of warfare or resource scarcity. Others collapsed quickly after European diseases tore through populations with unimaginable speed.

Yet for each city that fell, another rose elsewhere. Innovation did not die. Culture did not vanish. It moved, adapted, survived, and resurfaced in unexpected places.

The story of America's ancient cities is not one of disappearance alone, but of ongoing transformation.

## A Closing Invitation to the Reader

As you move into the first chapter, imagine the land beneath you, wherever you live, as a layered archive. Beneath pavement, riverbanks, fields, forests, and suburbs lies a world shaped by generations of people whose stories remain incomplete yet far from silent.

The vanished cities of the Americas do not fade when we stop looking at them. They remain embedded in soil, stone, memory, and the cultural traditions of descendant communities. Their echoes challenge familiar narratives. They remind us that history is broader than the timelines once written in textbooks, deeper than the ruins that survived erosion, and richer than what early archaeologists imagined.

The journey ahead will explore the cities that rose, flourished, and fell, and the legacies that remain. It will reveal how ingenuity and fragility coexist, how societies adapt to changing landscapes, and how understanding the past reshapes our view of the present. Step now into the quiet floodplain at dawn, where a mound becomes a map of memory, and let the first city show you the world that lies beneath the surface.

# Chapter 1
# The City Beneath the Cornfields — Rediscovering Cahokia's Forgotten Power

Sunrise comes slowly to the Mississippi River floodplain. A pale gold light rolls over wetlands and low grasses, catching the tops of reeds still heavy with dew. Mist clings to the open fields, drifting in thin veils across the gentle slope of an enormous earthen mound. Birds cut across the morning air and circle above a summit that once carried timber temples and symbols of power. This is Monks Mound, the largest pre-Columbian earthwork in the Americas, rising more than one hundred feet above the Illinois landscape. Today it sits quietly between highways, farmland, and subdivisions. Yet beneath these ordinary acres lies the footprint of a vanished metropolis.

As the camera pulls back in the mind's eye, the patchwork of cornfields and modern streets seems to dissolve. The clean lines of the present blur into the faint geometry of a forgotten world. Rectangular plazas appear where soybean rows now run straight. Wide causeways link ceremonial spaces. The arc of a wooden palisade encircles neighborhoods long since returned to soil. Modern noise fades. In its place comes the distant thunder of tens of thousands of footsteps across packed-earth streets. Fires crackle. Drums echo. Families climb steep stairways to ritual platforms. Priests and political leaders gather above plazas where crowds fill the open spaces. For a moment, Cahokia is alive again.

## A New Kind of City

Between roughly 1050 and 1350 CE, Cahokia became the largest city north of Mexico. Located near present-day Collinsville, Illinois, just east of St. Louis, it flourished in a fertile corridor where the Mississippi, Missouri, and Illinois Rivers converged. This crossroads provided transportation, abundant food sources, and access to distant trade routes. When Cahokia reached its height around 1100 CE, it may have held between fifteen and twenty thousand people within its core, with many more in the surrounding region. That population rivaled or exceeded medieval European cities such as London and Paris during the same era.

The city's sudden rise remains one of the most dramatic transformations in North American archaeology. Around 1050 CE, a surge of construction reshaped the landscape. Earth was moved on an immense scale, carried basket by basket and mound by mound. By the time Cahokia reached maturity, it contained more than one hundred earthen mounds of various sizes and shapes. Some rose in steep pyramidal forms for temples and ceremonial buildings. Others served as platforms for elite residences or burial structures. Monks Mound, the central monument, required an estimated twenty-two million cubic feet of soil carried from surrounding borrow pits. Its footprint covers more ground than many modern sports stadiums.

This extraordinary labor did not happen by accident. It required organized leadership, social coordination, and a belief system powerful enough to motivate monumental effort. Although the Mississippian world did not have writing, the archaeological record makes clear that Cahokia was a planned and deeply symbolic city. Streets and plazas aligned with celestial events. Mounds formed a landscape of power and meaning. Its political and religious center dominated a broad region, influencing outlying settlements across the American Midwest and Southeast.

## The Evidence Beneath the Ground

The rediscovery of Cahokia began long after the city itself had faded. By the nineteenth century, local farmers knew the mounds existed but misunderstood their origin. Some early settlers believed they were natural hills. Others assumed they were the work of a mysterious "lost race," a misconception that circulated widely before archaeology established their true Indigenous origins.

Modern excavation began to change everything. During the twentieth century, archaeologists uncovered remnants of neighborhoods, workshops, ceremonial spaces, and fortifications. One of the most striking discoveries came in the 1960s, when researcher Warren Wittry identified a circle of postholes northeast of Monks Mound. The pattern formed a large, circular calendar structure. Wittry called it Woodhenge, echoing its functional similarity to stone circles in Europe. The posts aligned with solstices and equinoxes, marking important moments in the agricultural and ceremonial calendar. Later excavations revealed multiple woodhenges built and rebuilt over generations.

These wooden calendars helped define Cahokia's social and spiritual life. They showed that astronomical knowledge played a central role in planning ceremonies, timing agricultural activities, and reinforcing political authority. Observing the rising sun from specific posts would have been a communal event that drew spectators together in large gatherings and reinforced the city's shared identity.

Other discoveries revealed a more sobering side of Cahokia's history. Excavations uncovered mass burials, some containing victims who showed signs of violent death. The most famous of these is found in Mound 72, where archaeologists uncovered more than two hundred and fifty individuals in multiple layers. Some were elite burials accompanied by grave goods, including finely crafted shell beads arranged in patterns. Others were sacrificial victims. Although the full meaning of these burials is still debated, they point to complex social and religious structures. Human sacrifice, though difficult to confront, appears to have been practiced in specific ceremonial or political contexts.

Craft production also flourished. Artifacts include shell ornaments from the Gulf Coast, copper from the Great Lakes, mica from the Appalachian Mountains, and finely made earthenware vessels. These items traveled along vast trade routes and show Cahokia's reach and influence. Skilled artisans shaped stone tools, pottery, figurines, and ritual objects. Their work reflects a society with specialized labor and elite-sponsored workshops.

## Life in a Mississippian Metropolis

Archaeological evidence suggests that Cahokia was more than a ceremonial center. It was a dense, vibrant community with neighborhoods, fields, plazas, and marketplaces. People lived in rectangular houses arranged around courtyards, forming social clusters of families or extended kin groups. Farming supported the population, especially maize agriculture, which expanded dramatically during Cahokia's rise. Maize provided a stable food supply and supported population growth. Farmers also cultivated squash, beans, and other crops, while wild foods such as deer, fish, nuts, and berries remained part of the diet.

This agricultural foundation helped fuel Cahokia's emergence as a political and ceremonial powerhouse. Leaders likely distributed food during large gatherings or seasonal events, strengthening their authority and reinforcing social bonds. The city's political structure was probably hierarchical, with elite rulers at the center. Their power was tied to religious authority, celestial alignments, and the ability to mobilize labor for public works.

Religion infused daily life. Mounds served as platforms for temples where rituals connected the earthly world with the sky realm and the underworld. Mississippian belief systems emphasized cycles of fertility, renewal, and cosmic balance. Public plazas hosted dances, feasts, and ceremonies that shaped the rhythm of life. Across the region, Cahokian symbols spread: bird motifs, hand-and-eye designs, copper falcons, and other imagery tied to warriors, leaders, and mythic beings.

## The Broader Mississippian World

Cahokia did not stand alone. Across the southeastern and midwestern United States, Mississippian communities shared similar artistic styles, agricultural systems, and mound-building traditions. Sites such as Moundville in Alabama, Etowah in Georgia, and Spiro in Oklahoma show cultural links to Cahokia's influence. While each had its own identity, they formed part of a broad cultural network connected through trade, ritual, and shared symbolism.

Cahokia's reach extended far beyond its immediate surroundings. Artifacts found at distant sites show the exchange of goods, ideas, and practices. The city may have influenced political structures in neighboring regions, especially through its ceremonial imagery and monumental architecture. Some archaeologists propose that Cahokia functioned as a regional center that exercised soft power through ritual prestige rather than through direct political control.

## Theories of Decline

By the mid-1300s, Cahokia had largely been abandoned. The reasons are complex, and no single explanation fully accounts for the city's decline. A combination of pressures appears to have converged.

Climate stress played a significant role. The Medieval Climate Anomaly, a period of warming, shifted toward the cooler and more volatile conditions of the Little Ice Age around the 1300s. These changes brought unpredictable rainfall, droughts, and flooding, all disruptive to an agricultural society dependent on maize. Archaeological evidence indicates cycles of major flooding in the Mississippi bottomlands, including events that may have damaged fields and settlements near Cahokia Creek.

Resource exhaustion is another factor. The city's rapid growth required timber for building, fuel, and defensive palisades. Over time, deforestation may have strained the landscape and made it harder to maintain the city's infrastructure. Soil depletion from intensive agriculture may also have reduced crop yields.

Social upheaval is harder to reconstruct but is suggested by changes in construction patterns and disruptions in trade. The city's elaborate mound-building slowed and eventually stopped. Evidence of increased fortification at certain stages suggests rising conflict or internal tension. Some scholars propose that political authority weakened, leading to population dispersal. Others note that religious or ceremonial practices appear to have changed or declined, indicating shifts in belief or leadership.

Disease is sometimes suggested, though there is no clear archaeological evidence for widespread epidemics at Cahokia. However, increasing population density often brings greater exposure to illness, even without the later devastations introduced through European contact.

Today the leading interpretation emphasizes a multi-layered collapse. Climate change, resource strain, political instability, and shifting social dynamics came together in a gradual decline rather than a single catastrophic event.

## Modern Rediscovery and Scientific Advances

The modern understanding of Cahokia owes much to decades of archaeological work. Scholars such as Warren Wittry, Melvin Fowler, and Timothy Pauketat transformed the site from a misunderstood mound complex into one of North America's most important ancient cities. Pauketat, in particular, has emphasized the scale and speed of Cahokia's rise, describing it as a cultural "Big Bang" that reshaped the region.

Scientific methods have deepened this picture. Radiocarbon dating refined the timeline of mound construction. Soil analysis revealed agricultural patterns. Isotope studies traced the origins of trade goods and identified where some of Cahokia's residents were born. Remote sensing technologies, including magnetometry and aerial LiDAR, mapped invisible structures beneath the surface without disturbing the ground. These tools revealed the outlines of houses, plazas, palisades, and ceremonial complexes now lost to time.

Such discoveries have changed how historians talk about early North American urbanism. For generations, many believed that Indigenous societies north of Mexico never developed large cities or centralized political structures. Cahokia challenges those assumptions. It reveals that urban life, monumental architecture, complex leadership, and long-distance trade networks existed in the heart of North America centuries before European arrival.

## Cahokia's Legacy

Cahokia's story continues to reshape public understanding of the continent's past. The city stands as evidence of Indigenous innovation, engineering, and social complexity at a scale long overlooked by mainstream history. Its influence echoes in the traditions of descendant communities across the Midwest and Southeast, including many Native nations whose heritage is tied to the Mississippian world.

Today the site is a UNESCO World Heritage Site, though much of ancient Cahokia remains hidden beneath modern development. Visitors climb Monks Mound, walk across the Grand Plaza, and visit reconstructed woodhenges that mark where ancient observers once watched the rising sun. Each structure connects past and present and shows how a vibrant city once thrived where interstate highways now intersect.

Cahokia's quiet fields offer a powerful reminder of how quickly cities can rise, reshape a landscape, and fade into obscurity. Its story fits within the broader theme of vanished urban worlds, places where human ambition pushed boundaries, built monumental structures, and shaped complex societies before yielding to forces they could not fully control. The ingenuity of Cahokia's builders, their ability to move millions of cubic feet of earth and organize a regional center of power, speaks to the depth of human creativity. Their decline, gradual and multi-causal, echoes across history and reminds readers that even the most impressive urban achievements remain vulnerable to environmental shifts and internal strains.

In the soft light of morning, with birds circling above fields that hide the traces of plazas and palisades, Cahokia invites viewers to reconsider what cities mean, how they form, and why they vanish. Its legacy endures not in stone or steel, but in the story it tells about the people who shaped North America long before the modern era and whose achievements continue to redefine the history of this land.

# Chapter 2
# Avenue of Shadows – Teotihuacan's Rise Without a Name

Dusk settles over the Basin of Mexico with a slow, fading glow. Along the Avenue of the Dead, long shadows stretch from the immense pyramids that flank the ancient roadway. Wind moves across dry grasses and abandoned plazas, brushing against stones that once felt the footsteps of tens of thousands of people. The Pyramid of the Sun looms ahead, its western face illuminated by the final light of the day. For a moment the structure flashes gold, then darkens again, leaving only the silhouette of a monument built to command an entire valley.

This silence is deceptive. Beneath the quiet surface lies the memory of what was once the largest metropolis in the ancient Americas, a city whose true name has been lost. Today it is known only by the name given to it by the Aztecs, Teotihuacan, “the place where gods were made.”

## The Making of a Metropolis

Teotihuacan did not begin as a city of giants. Its earliest occupation traces back to the first millennium BCE, when small farming communities settled near freshwater springs and fertile soils. Over time, these scattered villages merged and grew. A key moment came around 100 BCE, when a dramatic shift in political and economic organization triggered a wave of construction. The city’s planners reshaped the landscape into a precise grid aligned about fifteen degrees east of north. This alignment placed the main avenue and major monuments in harmony with astronomical cycles and nearby mountain peaks.

By the first few centuries CE, this careful planning reshaped the basin. Teotihuacan expanded into a massive urban center covering eight square miles, an extraordinary scale in the ancient world. At its height, between 150 and 550 CE, the city may have housed one hundred thousand to two hundred thousand people, a population far larger than most cities on Earth at the time.

The city’s design was intentional and symbolic. Streets and buildings followed strict orientations. The Avenue of the Dead ran for more than a mile, lined with ceremonial platforms and monumental structures. The Pyramid of the Sun and the Pyramid of the Moon anchored the urban landscape. To the southeast, the Ciudadela housed the Temple of the Feathered Serpent, decorated with stone carvings of snake and storm deities. These structures were not just civic works. They were expressions of cosmic order, political authority, and religious identity.

## A Multiethnic City

Archaeology reveals that Teotihuacan was a remarkably diverse city. Neighborhoods were organized into apartment compounds designed to house dozens of people. These compounds were built from stone and plaster and often decorated with vibrant murals that illustrated gods, rituals, and scenes from daily life. Standard features included central courtyards, altars, workshops, and living quarters. The presence of these apartments suggests strong social organization and a population accustomed to urban living.

Many residents came from far beyond the Basin of Mexico. Evidence of multiethnic communities has been found across the city. The Tlailotlacan barrio housed migrants from Oaxaca, identifiable by their pottery styles, burial practices, and distinctive artistic traditions. A neighborhood near the Ciudadela appears to have included people from the Gulf Coast, while other districts show cultural ties to the Maya world. These communities retained elements of their original identities while contributing to a shared urban culture.

Teotihuacan's diversity was not accidental. The city occupied a strategic position that allowed it to control trade routes across central Mexico. Its population included merchants, artisans, priests, laborers, and political elites, all intertwined within a complex social fabric.

## Engines of Economy and Craft

The city's economic strength rested on several key industries, especially the production of obsidian. The nearby Pachuca and Otumba sources provided high-quality volcanic glass that could be shaped into blades, spear points, and ritual objects. Teotihuacan became one of the main suppliers of obsidian throughout Mesoamerica. Workshops produced standardized tools at an industrial scale, helping the city maintain economic influence across distant regions.

Trade extended far beyond obsidian. Goods found at the site include shells from coastal areas, jade and greenstone from distant mountains, cacao from tropical forests, and ceramics reflecting artistic styles from across Mesoamerica. Teotihuacan was not only a marketplace but also a distributor of ideas. Its architectural styles, mural paintings, and religious imagery appear in far-flung locations, suggesting either strong cultural prestige or political reach.

Agriculture supported the population through irrigation canals, terrace systems, and carefully managed fields. Residents grew maize, beans, squash, chili peppers, and other crops. The surrounding wetlands offered additional resources, including fish, birds, and reeds used for weaving. The ability to maintain stable food supplies for such a large urban population demonstrates sophisticated environmental engineering.

## Engineering and Monumental Construction

The engineering behind Teotihuacan's monuments remains a major point of study. The Pyramid of the Sun, completed around 200 CE, rises more than two hundred feet and covers an area comparable to the Great Pyramid of Giza. Although its original exterior plaster has eroded, its shape and scale speak to extraordinary coordination and labor. Beneath the pyramid lies a large, man-made cave shaped into a four-leaf pattern. Some researchers believe this cave served as a place of origin in the city's mythology.

The Pyramid of the Moon anchors the northern end of the Avenue of the Dead. Archaeological excavations uncovered offerings beneath its construction layers, including obsidian blades, greenstone figurines, animals such as wolves and jaguars, and human sacrificial victims. These burials reflect a ceremonial role tied to warfare, rulership, and cosmic balance.

At the Ciudadela, the Temple of the Feathered Serpent represents another engineering feat. Its stone façade displays carved serpent heads surrounded by glyph-like ornaments. Beneath the structure, archaeologists uncovered more than two hundred human sacrifices arranged in mass burials, evidence of a city that intertwined religion and political authority with military power.

Water management was another cornerstone of Teotihuacan's design. The city's planners redirected rivers, built canals, and created drainage systems to control seasonal floods. These systems reflect an understanding of the Basin of Mexico's hydrology and an ability to harness natural forces for urban stability.

## Political Structure and Religious Life

One of the most intriguing mysteries surrounding Teotihuacan is its political organization. Unlike the Maya, who left written records of kings and dynasties, Teotihuacan produced no known written history. No royal tombs have been identified. Instead, many scholars argue that power may have been shared among councils or collective elite groups rather than centered on a single autocratic ruler. Monumental art focuses on deities and symbolic imagery rather than portraits of individuals, reinforcing the idea of a more collective leadership structure.

Religion permeated every part of the city. Murals found in apartment compounds display storm gods, fertility symbols, mythic animals, and scenes of communal rites. One of the most famous images, often called the Great Goddess, shows a towering figure with flowing water and plants emerging from her hands. This deity may have symbolized fertility, abundance, and the sustaining forces of nature. The city's alignment with mountains and celestial events suggests that urban planning itself was an act of devotion.

Rituals likely occurred in plazas that could hold thousands of people. Processions along the Avenue of the Dead would have reinforced civic identity. The presence of sacrificial victims in ceremonial contexts indicates that violence played a role in maintaining cosmic order, as it did in many Mesoamerican traditions.

## Collapse and Abandonment

By the sixth century, signs of decline became visible. Construction slowed. Trade networks weakened. Some neighborhoods experienced abandonment. Then came one of the most dramatic archaeological signatures at the site, the burning of elite compounds.

Excavations show that several major structures, including parts of the Ciudadela, were intentionally set on fire. Burn patterns indicate deliberate destruction rather than accidental damage. This burning has fueled theories of internal revolt, possibly by residents who turned against the ruling elite during a time of stress.

Climate change also played a significant role. Tree-ring studies indicate a prolonged drought in central Mexico during the sixth century. Crop failures and resource shortages may have strained the city's population. As food insecurity grew, the ability of leaders to command loyalty likely weakened. Combined with the outbreak of social unrest, the drought may have triggered a gradual withdrawal from the city.

By about 550 to 600 CE, Teotihuacan had lost much of its population. Smaller communities continued to occupy parts of the city, but its days as a major urban center had ended.

## The Aztecs and the Sacred City

Centuries later, when the Aztecs arrived in the Basin of Mexico, Teotihuacan was already an ancient ruin. Its origins were mysterious. Its monuments were towering reminders of a forgotten age. To the Aztecs, the city held sacred meaning. They believed it was the birthplace of the sun and the moon, the place where gods sacrificed themselves to set the cosmos in motion.

The Aztecs gave the city its present name and incorporated its symbolism into their own mythology. Pilgrims visited the ruins. Priests climbed the pyramids. Offerings were placed in chambers and caves. Teotihuacan's legacy shaped Aztec identity, even though the people who built it were long gone.

## Modern Archaeology and New Discoveries

Modern excavations continue to reveal how much remains hidden beneath Teotihuacan's surface. In recent decades, archaeologists have uncovered murals showing elaborate ritual scenes, workshops where artisans shaped obsidian tools, and apartment compounds that reveal details of everyday life.

One of the most remarkable discoveries came in the early 2000s, when a tunnel was found beneath the Temple of the Feathered Serpent. This tunnel contained thousands of artifacts, including jade, shells, ceramics, and carved stone. Pools of liquid mercury, symbolic of the underworld in some Mesoamerican beliefs, were found at its end. These discoveries suggest that the city's leaders staged elaborate rituals beneath the temple, connecting the earthly and cosmic realms.

Remote sensing technologies, including ground-penetrating radar and aerial mapping, continue to reveal buried plazas, compounds, and infrastructure. Each discovery expands our understanding of a city that left no written record yet shaped centuries of cultural history.

Teotihuacan's broad avenue and silent pyramids hold the memory of a city that once rivaled the great urban centers of the ancient world. Its rise demonstrates the power of planning, innovation, and collective identity. Its decline reveals how environmental stress, political turmoil, and internal conflict can undermine even the most impressive civilizations.

Like other vanished cities in this series, Teotihuacan stands as both marvel and warning. It shows what people can accomplish when they unite behind shared purpose, and how fragile even the greatest achievements can be. The city's shadowed monuments remind modern viewers that history is shaped not only by those whose names endure, but also by those who built extraordinary worlds without leaving their names at all.

# Chapter 3 Kingdoms in the Green — Maya Cities Lost to the Forest

The helicopter drops beneath the cloud line and skims low over a sea of green. Leaves brush the air in waves, their motion barely visible from above, but the sound, steady, alive, and relentless, vibrates through the humid atmosphere. The canopy spreads in every direction, unbroken and ancient. Then the trees part. Out of the jungle rise stepped pyramids of pale limestone, their edges softened by centuries of vines and rain. Terraces appear beneath them, descending into plazas where royal courts once gathered under banners of bright feathers and incense smoke. The air carries the buzz of insects, the cries of distant birds, and the low rumble of thunder rolling across the horizon. These are the lost kingdoms of the Maya, once powerful city-states that dominated the rainforests of Mesoamerica.

## Cities of Kings and Calendars

From about 250 to 900 CE, dozens of Maya cities flourished across what is now Guatemala, Mexico, Belize, Honduras, and El Salvador. Each city-state operated as an independent kingdom ruled by a dynastic line of kings, queens, and noble families. Tikal, one of the largest and most influential cities, rose above the lowland jungle with towering pyramids that reached more than two hundred feet. Its Great Plaza formed the ceremonial heart of a vast urban landscape connected by causeways, reservoirs, and residential compounds.

Palenque, set in the foothills of Chiapas, became known for its elegant palaces, intricately carved stucco, and a royal lineage that included the famous ruler K'inich Janaab' Pakal. His tomb, discovered in 1952 within the Temple of the Inscriptions, offered one of the most extraordinary archaeological finds in the Americas, a richly decorated sarcophagus surrounded by carved glyphs recording his reign.

Far to the south, Copán thrived in the Motagua River valley. Renowned for its stelae, tall sculpted monuments depicting kings in elaborate regalia, Copán developed a sophisticated school of astronomy and mathematics. One ruler, known as 18 Rabbit, commissioned some of the finest stone carvings in Maya history.

Calakmul, deep within the forests of the Yucatán Peninsula, rivaled Tikal in size and political power. Its rulers, part of the powerful Snake Dynasty, forged alliances and influenced dozens of smaller kingdoms. For centuries, Calakmul and Tikal engaged in a long struggle for dominance, a rivalry recorded in glyphs carved into stone monuments across the region.

## Architecture of Power and Ritual

Maya cities were built to impress both subjects and the gods. Their most visible structures, pyramids, temples, and palaces, rose above broad plazas designed for public ceremonies. Many of these buildings also served as astronomical observatories. Windows, doorways, and staircases aligned with solstices, equinoxes, and the movements of Venus, a planet central to Maya cosmology and the timing of warfare.

Beneath the towering structures lay complex systems of reservoirs, canals, and terraces. Many cities lacked permanent rivers, so residents engineered ways to capture and store rainwater. Tikal, for example, built a network of catchment basins that sustained tens of thousands of people through dry seasons. These reservoirs were lined with clay and maintained with careful precision.

The Maya writing system, one of the most advanced in the ancient world, recorded the lives of rulers, battles, marriages, rituals, and dynastic claims. Hieroglyphs appeared on stone stelae, painted codices, ceramics, and the walls of temples. The script combined phonetic signs and logograms, allowing scribes to record rich narratives tied to specific dates. Maya mathematics and calendrical systems reached remarkable levels of accuracy. They tracked lunar cycles, solar years, and the long count of days that stretched back to mythic creation.

## A Network of Allies and Rivals

Maya cities did not exist in isolation. They formed a network of alliances, marriages, trade relationships, and military rivalries that shaped regional politics. Diplomacy and war were central to Maya statecraft.

Tikal and Calakmul exemplify these dynamics. Their rivalry spanned centuries, shifting the political landscape as each sought allies among smaller kingdoms. These alliances often shifted based on marriages, access to trade routes, or military support. When one city gained ascendancy, the other responded with strategic campaigns or diplomatic maneuvers.

Trade routes carried obsidian from highland Guatemala, jade from the Motagua Valley, salt from coastal flats, and cacao from tropical regions. Materials and artistic styles traveled widely, linking distant communities under shared cultural traditions. Maya elites used these goods to display status, strengthen political ties, and reinforce religious authority.

## Life in the Rainforest Kingdoms

Archaeology reveals that Maya cities were dense, vibrant places. Residential zones included houses made of perishable materials built atop low platforms. Markets sold food, pottery, textiles, and luxury goods. Specialists crafted jade ornaments, ceramic vessels, painted murals, and stone tools. Farmers cultivated maize, beans, squash, cacao, cotton, and tropical fruits. Terracing and raised-field agriculture helped manage soil depletion and seasonal flooding.

Religion shaped daily life. Temples served as portals between the human world and the supernatural realm. Rituals included offerings, feasting, incense burning, music, and, in some cases, human sacrifice. The Maya believed that rulers mediated between the earthly and divine spheres. Their authority depended on their ability to perform rituals, conduct warfare successfully, and maintain cosmic harmony.

## The Classic Maya Collapse

Around the ninth century, many major Maya cities experienced a decline so dramatic that it became known as the Classic Maya Collapse. The process was neither sudden nor uniform. Some cities fell earlier, others survived longer, and the causes varied. Yet several broad forces stand out.

Drought cycles played a pivotal role. Sediment cores from lakes across the Maya region show repeated severe droughts during the eighth and ninth centuries. These droughts reduced water supplies and agricultural productivity, placing enormous strain on cities reliant on reservoirs.

Warfare intensified as resources dwindled. Archaeological evidence from sites like Dos Pilas shows fortified walls and signs of violent conflict. Stelae record battles, captured rulers, and shifting alliances. As rivalries escalated, many kingdoms were destabilized.

Political fragmentation followed. As drought, warfare, and internal pressures mounted, the authority of divine kings weakened. In some cities, monuments ceased to be erected. Others show signs of abrupt abandonment.

Ecological pressure contributed to the crisis. Deforestation for fuel, construction, and agriculture likely worsened soil erosion and disrupted local climates. Pollen records indicate that forest loss intensified in key regions and reduced resilience to drought.

Shifting trade networks affected economic stability. As coastal and highland routes changed, some inland cities lost access to valuable goods, further undermining elite power.

The collapse was not total. Northern cities such as Chichén Itzá continued to flourish after the southern lowland kingdoms declined. Many Maya communities survived, adapted, and persisted into the Postclassic era and beyond. Today millions of Maya people continue cultural traditions that stretch back thousands of years.

## Archaeological Breakthroughs

The modern understanding of the Maya owes much to extraordinary advances in archaeology and technology. Two developments stand out, the decipherment of Maya hieroglyphs and the use of LiDAR scanning.

The decipherment of the script began in the mid-twentieth century, with key breakthroughs by researchers such as Tatiana Proskouriakoff and Yuri

Knorosov. Their work revealed that the glyphs recorded historical events rather than only religious or mythological narratives. Suddenly, stelae and murals became readable. Scholars could identify rulers, dynasties, battles, and alliances. This revolution transformed Maya archaeology from the study of architecture and artifacts into the study of a literate civilization with rich historical records.

LiDAR, light detection and ranging, has further reshaped the field. By sending laser pulses through dense vegetation, LiDAR can map the ground surface beneath the canopy with notable precision. Since 2018, LiDAR surveys have uncovered entire landscapes of hidden structures, causeways linking distant cities, vast agricultural terraces, defensive walls, and thousands of unknown buildings. In Guatemala's Petén region alone, LiDAR revealed more than sixty thousand previously unidentified structures, demonstrating that Maya civilization was far more interconnected and populous than once believed.

These discoveries show that even well-known cities such as Tikal and Calakmul were part of a larger, complex network of settlements integrated by roads, waterways, and shared political interests.

## The Forest as Guardian and Obscurer

For centuries after their decline, Maya cities were reclaimed by the rainforest. Trees rooted themselves in terraces and temple platforms. Vines covered facades once carved with detailed imagery. Structures collapsed under their own weight or were buried in soil washed down from surrounding hillsides. In many areas, the forest preserved what it concealed. Limestone buildings protected under layers of earth remained intact until archaeologists exposed them.

This dual role of the rainforest, destroyer and preserver, allowed many Maya cities to survive long enough to be rediscovered. Early explorers in the nineteenth century found buildings emerging from dense foliage. Since then, archaeological excavation, careful conservation, and modern technology have revealed more of these kingdoms each decade.

Maya cities stand today as silent monuments within the forest, their towers rising above green waves like islands in a living sea. Their rise demonstrates exceptional achievements in astronomy, engineering, water management, writing, and governance. Their decline reminds us how complex societies can unravel when environmental stress collides with political fragility.

Yet these kingdoms were never truly lost. They continued through the people who lived on, adapted, and preserved traditions that endure today. As LiDAR reveals vast networks beneath the canopy and glyphs continue to illuminate ancient political histories, the Maya world grows ever clearer. These cities, once hidden beneath the forest, reemerge as powerful reminders of human ingenuity and the delicate balance between civilization and its environment.

# Chapter 4
# The Empire That Vanished Twice — Tiwanaku and the Mystery of Lake Titicaca

Storm clouds sweep across the vast altiplano, dragging shadows over the cold waters of Lake Titicaca. Wind whistles across open plains that stretch for miles in every direction, carrying the scent of rain and thin mountain air. At 12,500 feet above sea level, every breath feels deliberate, and every footstep echoes sharply through ancient courtyards. In the shifting light, stone gateways rise from the ground like frozen sentinels. Carved monoliths stand silently beside them, their surfaces etched with patterns that catch faint glimmers of sun. This is Tiwanaku, a place shaped by brilliant engineering, complex ritual, and a mystery of rise and disappearance that unfolded twice across Andean history.

## The Birth of a High-Altitude Power

Long before the Inca built their empire, the people of Tiwanaku established a thriving civilization on the southern shores of Lake Titicaca. The settlement began centuries earlier, but between 500 and 1000 CE it transformed into a major urban and ceremonial center. Its location might seem unlikely. The region experiences freezing nights, strong winds, and thin air. Yet the lake moderates the climate enough to allow crops to grow, and the surrounding plains offered space for innovation on a grand scale.

Tiwanaku's success was rooted in its mastery of agriculture at high altitude. Farmers created raised-field systems known as sukakollos, which combined elevated planting surfaces with water-filled channels. These channels absorbed heat during the day and released it at night, protecting crops from frost. The system boosted yields of potatoes, quinoa, and other Andean staples. This innovation strengthened food security and supported a growing population.

As agricultural output increased, Tiwanaku expanded its influence across the Andean highlands. The city became a center of trade, ritual authority, and political organization. By 800 CE, it controlled or influenced communities stretching from modern Bolivia into Peru and northern Chile. Its power rested not only on force, but also on religion, agriculture, and cultural prestige.

## Stone Engineering and Monumental Spaces

Tiwanaku is best known for its extraordinary stone architecture. Visitors today walk through ceremonial complexes built with precision-cut blocks of andesite and sandstone, some weighing many tons. The stones fit so tightly that a knife blade cannot pass between them. How these stones were transported and shaped remains a topic of research, but their craftsmanship signals a highly organized workforce and specialized knowledge.

The most iconic structure is the Gateway of the Sun, a single block of andesite carved with intricate images linked to Andean deities and celestial cycles. Its central figure, often called the Staff God, appears in many regional cultures and suggests that Tiwanaku served as a spiritual hub for surrounding communities.

Nearby stands the Kalasasaya platform, a vast raised plaza framed by tall upright stones. It likely functioned as an astronomical observatory, aligned with solstices and seasonal events. During rituals, crowds may have gathered here to watch priests interpret the sky, reinforcing Tiwanaku's authority as both a political and cosmic center.

Below the Kalasasaya lies the Semi-Subterranean Temple, a sunken courtyard lined with carved stone heads set into its walls. Some faces resemble local styles, while others appear foreign, possibly representing the diversity of peoples connected to Tiwanaku's influence.

Across the site, monoliths depict human-like figures holding ceremonial objects, wearing elaborate clothing, and conveying both power and solemnity. These carvings are among the finest in Andean art, capturing a worldview rich with symbolism, ritual complexity, and attention to detail.

## Networks of Influence

Although Tiwanaku lacked writing, its reach is visible through material culture. Pottery styles, textile patterns, stone carvings, and ritual imagery appear in regions far beyond its core territory. These markers suggest a network of trade, pilgrimage, and political alliances.

Tiwanaku elites likely traveled widely, forging relationships through marriages, ceremonies, and the exchange of goods. Copper, textiles, obsidian, llama caravans, and food staples moved along mountain routes connecting communities across the Andes. Colonies and partner settlements, sometimes referred to as enclaves, helped extend Tiwanaku's presence into valleys and coastal regions.

Art played a crucial role in unifying this cultural sphere. Iconography linked to the Staff God and other deities appears throughout the region, indicating a shared religious language. Ceremonies associated with agricultural renewal, ancestor veneration, and water management reinforced Tiwanaku's spiritual authority.

## Water, Ritual, and the Sacred Landscape

Water shaped Tiwanaku's identity. Lake Titicaca, one of the highest navigable lakes in the world, was central to Andean cosmology. Its shores inspired creation myths, including legends that later shaped Inca beliefs. To the Tiwanaku civilization, the lake represented both life and mystery, especially in a region where rainfall could be unpredictable.

Rituals involving water appear throughout the site. Channels, drains, and sunken courts directed rainwater into symbolic spaces. Some monuments may have been designed to interact with reflective pools, blending water, architecture, and celestial observation.

The surrounding landscape also held sacred significance. Mountains, considered living beings in Andean tradition, formed the horizon and shaped ritual geography. Tiwanaku's builders integrated the city with this larger environment, aligning structures with important peaks and celestial events.

## The Enigma of Abandonment

Despite its strength, Tiwanaku began to weaken around 1000 CE. Unlike cities destroyed by conquest or volcanic eruption, its decline appears to have been gradual. Archaeological evidence suggests that environmental change played a major role.

Lake Titicaca's water levels fluctuated dramatically during this period. Droughts became more frequent. The raised-field agricultural system, dependent on consistent water availability, suffered as the climate shifted. Studies of lake sediments reveal reduced rainfall and prolonged dry cycles, disrupting crop production across the region.

With agricultural yields declining, Tiwanaku likely faced food shortages. Its population began to disperse, and the political structure that once held the region together fractured. Ritual centers fell into disuse. Workshops closed. Over time, the city emptied, leaving behind stone monuments exposed to wind, rain, and shifting soils.

By the time the Inca expanded into the area centuries later, Tiwanaku was already an ancient ruin. Yet its cultural memory endured. The Inca held Tiwanaku in high esteem, viewing it as a place connected to their own origins. They incorporated elements of Tiwanaku's art and architecture into their imperial ideology, and Inca rulers made pilgrimages to the site.

## Rediscovery and Archaeological Advances

Modern archaeology has brought Tiwanaku's achievements back into focus. In the twentieth century, scholars such as Max Uhle, Wendell Bennett, and later Carlos Ponce Sanginés helped establish a clearer timeline for the site. Excavations uncovered temples, elite residences, ceremonial platforms, and workshops where artisans shaped stone tools and ceramics.

Recent advances have pushed research even further. Underwater surveys of Lake Titicaca have revealed ritual offerings, including carved stone boxes, gold ornaments, and ceramic vessels placed near submerged islands. These discoveries demonstrate the lake's role as a sacred space and deepen understanding of Tiwanaku's ceremonial life.

Ground-penetrating radar and digital mapping have uncovered buried terraces, canals, and causeways. These tools confirm that Tiwanaku was not just a single city, but part of a larger urban and agricultural landscape integrated into the surrounding altiplano.

Bioarchaeological studies offer insight into the lives of Tiwanaku's people. Analyses of bones and teeth reveal diverse diets, long-distance migration patterns, and the presence of individuals from various regions of the Andes. These findings support the view that Tiwanaku's influence extended far beyond its immediate surroundings.

## The Legacy That Shaped Empires

Although Tiwanaku fell centuries before the rise of the Inca Empire, its legacy influenced the Andean world in profound ways. The Inca adopted elements of Tiwanaku's statecraft, religious imagery, and agricultural techniques. Raised-field agriculture and terracing became essential to Inca food production. Iconography linked to the Staff God continued to appear in Inca art. Even the Inca origin myths, many of which describe ancestors emerging from Lake Titicaca, reflect Tiwanaku's earlier cultural memory.

Tiwanaku also provided a model for how a high-altitude society could thrive. Its mastery of frost-resistant agriculture, water management, and monumental architecture demonstrated what was possible in the challenging environment of the altiplano. These lessons helped shape later Andean civilizations and remain relevant for modern researchers studying climate resilience.

Today Tiwanaku stands quiet under the sweeping skies of the Bolivian highlands, its stone monuments echoing faintly with the footsteps of visitors. Storm clouds still gather over Lake Titicaca, casting shifting light across gateways and monoliths that once framed ceremonies and cosmic observations. Though abandoned a millennium ago, Tiwanaku remains a testament to human innovation in one of the world's harshest environments.

Its rise reveals how a civilization can transform landscape into opportunity through engineering, agriculture, and collective belief. Its decline shows how fragile even the most sophisticated systems become when climate turns against them. Yet Tiwanaku's influence endured, absorbed into Andean memory, reflected in Inca tradition, and rediscovered through modern science.

As the winds sweep across the altiplano and thunder rolls across the lake, Tiwanaku reminds us that empires can vanish, reemerge in new forms, and vanish again, leaving behind stones that continue to speak across the centuries.

# Chapter 5
# Desert Giants — The Nazca Cities and the Lines No One Saw

The camera soars over a vast, barren plateau in southern Peru, where the midday sun casts sharp shadows across a landscape of wind-scoured earth. From above, shapes begin to appear: a hummingbird stretching nearly three hundred feet, a monkey with a spiraling tail, geometric trapezoids running for miles, and straight lines etched with precise clarity into the desert floor. The Nazca Lines spread across the pampa like a silent message intended for no human eyes on the ground. In the far distance, thin green ribbons of river valleys cut through the desert, where faint outlines of ancient settlements cling to the margins of life. These valleys once supported communities that built ceremonial centers, irrigated gardens, and crafted the cultural world that gave rise to the famous geoglyphs.

## Cities of the Desert Fringe

The Nazca civilization flourished between about 200 BCE and 600 CE along the dry coastal plains of southern Peru. Unlike the highland cultures of the Andes, the Nazca built their world in one of Earth's driest regions, where survival required careful management of water. Their settlements developed in small clusters along river valleys fed by seasonal rains and underground aquifers. Life centered on these narrow corridors of fertility, where crops such as maize, beans, cotton, and squash could grow.

Cahuachi became the most important ceremonial center of the Nazca world. Located near the Nazca River, it functioned not as a typical urban city but as a pilgrimage and ritual complex. Spread across more than a square mile, Cahuachi included stepped platforms, adobe temples, plazas, and ritual storage areas. Its architecture relied on earth and adobe rather than stone, which made it vulnerable to erosion yet still impressive in scale.

Ceremonies at Cahuachi were tied to agricultural cycles, water availability, and community identity. Thousands of people likely gathered there during major festivals, bringing offerings, textiles, decorated pottery, and food. For centuries, Cahuachi served as a unifying site for dispersed settlements throughout the region and reinforced shared beliefs and cultural cohesion.

## Irrigation and the Struggle for Water

Water shaped every aspect of Nazca life. The region receives almost no rainfall, so the Nazca engineered systems to capture and transport underground water. They built puquios, spiral-shaped access channels leading to subterranean aqueducts. These aqueducts tapped into natural groundwater flows and delivered water to fields and settlements. Some puquios remain functional today, demonstrating the sophistication of Nazca hydrological engineering.

Canals extended across the desert margins, feeding agricultural terraces and gardens. The control of water was not just practical, it was sacred. Rituals involving offerings, dances, and symbolic objects sought to ensure the return of seasonal rains and the health of irrigated fields. Water deities and fertility symbols appear prominently in Nazca art and reflect both fear and reverence for this scarce resource.

## Textiles, Pottery, and Artistic Innovation

The Nazca are widely known for their artistic achievements, full of color and intricate imagery. Their textiles, some preserved in the dry desert environment, feature elaborate embroidery and dyed fibers in vibrant shades of red, yellow, purple, and blue. The motifs include mythical beings, plants, animals, and supernatural creatures with flowing hair, fanged mouths, or elaborate headdresses. Such textiles likely played significant roles in ceremonies, social status, and offerings.

Nazca pottery is equally distinctive. Ceramic vessels display complex scenes painted in multiple colors, often depicting rituals, agriculture, and mythological figures. The iconography reflects a world in which humans, animals, and supernatural beings interacted through ceremonies tied to water, fertility, and cosmic forces.

These artistic traditions connect directly to the Nazca Lines. Many of the animals and geometric shapes found on pottery appear mirrored in the geoglyphs etched across the plateau. The imagery reinforces the idea that the lines were part of a larger cultural system grounded in ritual meaning and agricultural necessity.

## The Nazca Lines and Their Purpose

The Nazca Lines remain one of the most famous archaeological mysteries. The geoglyphs include hundreds of straight lines, geometric shapes, and animal figures stretching across more than two hundred square miles. Created by removing the sun-darkened stones on the surface to reveal lighter sediment beneath, the lines are remarkably well preserved due to the desert's lack of rain and wind.

The most widely accepted interpretation links the lines to water rituals and fertility practices. Many scholars propose that the lines functioned as ceremonial pathways. Groups may have walked the lines during rituals intended to honor deities connected with rainfall, groundwater, and agricultural cycles. The trapezoids and long straight lines may have marked pilgrimage routes or framed ritual movement across the pampa.

Astronomical interpretations also exist. Early researchers such as Paul Kosok and Maria Reiche proposed that some lines align with solstice events or star positions. While some alignments appear intentional, most archaeologists now argue that astronomical explanations alone cannot account for the full complexity of the geoglyph system.

Instead, the lines likely represent a blend of religious symbolism, ritual performance, and regional identity. Their immense scale suggests coordination among multiple settlements and participation in shared ceremonies that reinforced cultural unity.

## Collapse in the Desert

Around 500 to 600 CE, the Nazca culture experienced significant decline. Several factors contributed to this collapse, each rooted in the fragility of life on the desert's edge.

Prolonged drought is one major theory. Geological and botanical evidence indicates that climate shifts affected rainfall patterns across the Andes. With less water flowing through rivers and aquifers, agriculture suffered. As food production declined, populations gradually dispersed or turned to alternative subsistence strategies.

Deforestation also played a role, especially along river valleys where the Nazca cleared forests to plant crops. The removal of native huarango trees, deep-rooted species that stabilized soil and maintained groundwater levels, made the land more vulnerable to erosion. Without these trees, flash floods and shifting river courses began to wash away fertile soil.

Shifting river patterns combined with drought could have destroyed farmland and settlements alike. As riverbeds meandered or dried, irrigation systems failed, and communities no longer had reliable access to water. Archaeological evidence shows that Cahuachi was gradually abandoned around this time, signaling a breakdown in the ceremonial and social systems that held the culture together.

Environmental pressures collided with cultural transformations, leading to the dispersal of populations and the end of centralized Nazca ceremonial centers.

## Rediscovery and Changing Interpretations

The Nazca Lines remained virtually unknown to the wider world until the twentieth century. Local residents had long been aware of the lines at ground level, but their full scale became clear only when aircraft began flying over the region in the 1920s. Pilots noticed enormous shapes stretching across the desert, appearing as if carved by giant hands.

Early interpretations varied widely. Some speculated that the lines were astronomical calendars. Others imagined they were markers for ancient roads or symbols intended for deities in the sky. Some theories drifted into the fantastical, suggesting connections to extraterrestrial visitors, ideas that overshadowed the real cultural achievements of the Nazca themselves.

The work of researchers such as Maria Reiche helped shift scholarly and public understanding. Reiche devoted her life to studying and protecting the lines, arguing for their astronomical and ceremonial importance. Later archaeologists expanded the interpretation, connecting the lines to Cahuachi, water rituals, and wider ceremonial landscapes.

Recent studies using drone surveys, satellite imagery, and digital mapping continue to uncover previously unknown geoglyphs. These technologies show that the Nazca world was more interconnected than once believed, with lines extending far beyond the famous clusters on the central pampa.

## Cities Hidden in Plain Sight

Although the Nazca Lines dominate public imagination, the settlements that supported this culture were equally important. Excavations near the Nazca and Ica Valleys reveal residential compounds, workshops, cemeteries, and agricultural fields. The desert preserved textiles and pottery in remarkable condition, offering a clearer picture of everyday life.

Cahuachi's adobe temples still rise above their surroundings, though weathered by centuries of erosion. Ongoing research continues to uncover ceremonial platforms, offerings, and evidence of large gatherings. The surrounding fields show the remnants of irrigation canals and puquios, linking agriculture directly to the rituals performed at the site.

These connections reveal that the Nazca Lines were not isolated creations. They were part of a broader cultural landscape of cities, farms, trade networks, and ceremonial centers bound together by the constant pursuit of water and agricultural stability.

From the air, the Nazca Lines appear as messages etched into a barren world. On the ground, the desert seems empty, silent, and endless. Yet beneath this stillness lies the memory of a civilization that transformed one of Earth's harshest environments into a place of vibrant creativity and ritual meaning.

The Nazca engineered irrigation systems that still function today, created textiles and pottery rich in symbolic detail, and built ceremonial centers that anchored cultural identity. Their lines on the desert floor were part of a wider network of belief and survival. The culture's decline underscores the vulnerability of societies that depend on delicate environmental balances.

As new technologies continue to reveal hidden geoglyphs and forgotten settlements, the Nazca world grows more intricate. What once seemed mysterious becomes part of a coherent story of ingenuity, adaptation, and the enduring challenge of living in the desert. These giant drawings, visible only from above, remind modern viewers that some of humanity's greatest works were crafted by cultures whose lives depended on forces they could not fully control and whose legacy endures long after their cities faded into the sand.

# Chapter 6
# The Cliffside Exodus – Pueblo Cities of the Four Corners

Canyon walls glow deep orange in the fading light, their surfaces catching the sun's final warmth before plunging into evening shadow. High above the canyon floor, stone dwellings cling to the cliffs like honeycombs carved into the sandstone itself. Small windows and doorways stare outward from alcoves, their edges smoothed by centuries of wind. A soft whistle moves through the openings and echoes faintly across the chasm. This is the homeland of the ancestral Pueblo peoples, a region where ingenuity met harsh landscapes and where entire cities once flourished on mesas, canyon rims, and cliff faces throughout the Four Corners.

## Cities of Stone and Sun

Between about 900 and 1300 CE, the ancestral Pueblo world, stretching across present-day Colorado, Utah, New Mexico, and Arizona, supported a network of increasingly complex settlements. These communities built stone villages, engineered water systems, and developed ceremonial structures that reflected a deep connection to the land and sky.

Mesa Verde, perched atop high mesas in southwestern Colorado, is best known for its dramatic cliff dwellings. Structures such as Cliff Palace, Balcony House, and Spruce Tree House display masterful masonry. Builders shaped sandstone blocks into walls several stories high, fitted with careful precision, and arranged around plazas and ceremonial rooms. These cliff dwellings were not isolated. They were connected to earlier mesa-top villages through pathways, farming terraces, and reservoirs that captured precious seasonal water.

Chaco Canyon, in northwest New Mexico, served as a major ceremonial and administrative center for the wider region. Between 850 and 1150 CE, its inhabitants constructed monumental great houses using millions of shaped sandstone blocks and massive timber beams transported from distant mountain ranges. At its height, Chaco featured multi-story buildings with hundreds of rooms, internal plazas, and large circular subterranean chambers known as kivas.

Structures such as Pueblo Bonito, Chetro Ketl, and Kin Kletso were not ordinary residences. Their size, alignment, and layout suggest ceremonial, ritual, and political significance. Chaco's architects oriented many buildings along solar and lunar sightlines. Windows and doorways captured light during solstices or lunar standstill events, indicating an advanced understanding of astronomy.

To the north, Salmon Ruins near the San Juan River extended Chaco's cultural influence into what is now New Mexico. Its builders reproduced Chacoan architectural styles, including core-and-veneer walls and large kivas, signaling strong cultural connections. Salmon Ruins later evolved through migration and adaptation and linked Chacoan traditions with post-Chaco communities.

## Roads, Trade, and a Connected World

One of the most remarkable aspects of Chacoan society was its network of engineered roads stretching across hundreds of miles. These roads, some extending in straight lines for more than forty miles, linked Chaco Canyon to outlying settlements, great houses, and resource areas. Their construction required significant labor and coordination. Archaeologists have identified ramps, berms, stairways carved into cliffs, and road segments deliberately cleared of vegetation.

The road system facilitated more than transportation. It reinforced Chaco's regional authority and created a sense of shared identity among distant communities. Pilgrims, traders, and messengers likely used these routes to move goods and participate in ceremonies at Chaco's central plazas.

Trade extended far beyond the Four Corners. Excavations reveal turquoise from southwestern mines, seashells from the Pacific and Gulf of California, parrots and macaws from Mesoamerica, copper bells from West Mexico, and cacao residue in ceramic vessels. These materials show that ancestral Pueblo peoples engaged in long-distance exchange that reached deep into Mexico's cultural spheres.

Turquoise workshops at Chaco Canyon produced finely crafted ornaments used in ceremonies or exchanged as status symbols. A complex economy evolved around these trade networks and linked communities through shared ritual practices and resource distribution.

## Kivas and Ceremonial Life

Ceremonial architecture defined much of the ancestral Pueblo world. Kivas, circular subterranean rooms, served as spaces for rituals, governance, and community gatherings. Their design varied across regions, but many included fire pits, benches, ventilation shafts, roof openings, and symbolic features representing connections to the underworld.

In Chaco Canyon, massive great kivas measuring more than sixty feet in diameter required enormous logistical planning. These structures were likely used for ceremonies involving multiple communities, reinforcing regional cohesion. Smaller kivas within great houses served different purposes, possibly separating everyday rituals from major public events.

At Mesa Verde, kivas appeared in both cliff dwellings and earlier mesa-top villages. Their consistency across centuries reflects long-standing ceremonial traditions still recognized by Pueblo descendants today.

## Drought, Social Change, and Migration

The ancestral Pueblo world experienced significant population growth between 1000 and 1200 CE. Villages expanded, agricultural fields spread across mesas and canyon bottoms, and regional connections deepened. Yet this prosperity depended on a delicate environmental balance.

Tree-ring studies, a field known as dendrochronology, provide one of the clearest records of climate shifts in the American Southwest. The rings preserved in ancient timbers show patterns of precipitation, temperature, and environmental stress. These records indicate that severe droughts struck the region during the mid-1100s and again in the late 1200s.

As rainfall declined, cornfields failed, springs dried up, and communities faced growing pressure on resources. Archaeological evidence suggests increased social tension during this period. Defensive sites built on high mesas or protected by cliffs became more common. Some villages constructed watchtowers or enclosed plazas, hinting at conflict or fear of conflict.

Chaco Canyon began to decline in the 1100s. Construction slowed, and populations dispersed to nearby regions such as the Chuska Mountains and the Middle San Juan. Salmon Ruins shows signs of burning during a period of upheaval, though the causes remain debated.

By the late 1200s, the Mesa Verde region experienced widespread abandonment. Entire villages relocated south toward areas with more reliable rainfall and larger river systems, including the Rio Grande Valley and the Hopi mesas. This migration was not a collapse but a relocation, a major shift in settlement patterns that shaped the cultural landscape of the American Southwest.

## Archaeology and Oral Histories

Scientific archaeology has uncovered much of what is known about ancestral Pueblo cities, but descendant communities have preserved knowledge as well. Pueblo peoples today, including Hopi, Zuni, and various groups along the Rio Grande, maintain oral histories describing origins, migrations, and ancestral connections to places such as Chaco and Mesa Verde.

These stories, though not framed as archaeological narratives, reflect deep ties to landscape, ceremony, and community identity. They describe migrations undertaken in response to environmental changes, spiritual guidance, or social pressures. Such histories complement archaeological evidence and offer perspectives grounded in living cultural memory.

Excavations across the Southwest have revealed details of everyday life. Ceramic styles indicate changing trade patterns. Middens and food remains show shifts in diet and agriculture. Architecture reveals evolving social structures and responses to environmental stress. Tree-ring dating pinpoints construction timelines with precision, allowing researchers to correlate building activity with climate data.

At Mesa Verde, studies of plastered walls and soot patterns inside cliff dwellings reveal how families organized domestic space. In Chaco Canyon, isotopic analysis of timbers shows that many construction beams came from mountain ranges more than fifty miles away, demonstrating far-reaching resource networks.

## A Landscape of Persistence

Although ancestral Pueblo cities were largely abandoned by the end of the 1200s, their legacy remains visible across the Southwest. Later communities adapted earlier traditions, building new villages along rivers, mesas, and high desert plains. Ceremonial practices, agricultural knowledge, and architectural styles continued in modified forms.

Modern Pueblo nations maintain cultural ties to the Four Corners region. Many visit ancestral sites for ceremony, research, or cultural preservation. These places remain active components of identity, not frozen relics of the past.

As the sun sets over the canyon walls, the cliff dwellings of Mesa Verde glow softly against the sandstone. Their silent doorways face outward toward a landscape that once supported interconnected cities, regional roads, astronomical observatories, and flourishing trade networks. The ancestral Pueblo world, shaped by ingenuity and challenged by climate, demonstrated resilience across centuries.

The widespread migration of the late 1200s reminds modern readers that civilizations adapt when conditions change, even when the shift requires leaving behind monumental architecture and long-established homelands. What remains today is more than stone. It is a testament to a living culture whose descendants still honor these places and preserve traditions shaped by both struggle and innovation.

The wind continues to whistle through ancient doorways, carrying echoes of communities who built in partnership with the land and sky. Their story, written in sandstone and memory, endures as one of the most remarkable chapters in the history of the American Southwest.

# Chapter 7
# The River Kingdom — The Moche Cities and Their Sudden Silence

Eroded adobe pyramids rise from Peru's northern desert, their massive forms shaped by centuries of wind and rain. In the dim light of an approaching storm, the desert appears both generous and unforgiving. Low clouds gather above the river valleys that once sustained vibrant cities. Shadows fall across the remnants of temples, plazas, and ancient canal systems. This is the land of the Moche, a civilization that harnessed fragile watercourses to build thriving kingdoms, only to face cycles of catastrophe that unsettled their world and left behind a sudden silence across the northern coast.

## A Civilization Built on Rivers

Between about 100 and 800 CE, the Moche flourished along the narrow river valleys of Peru's north coast. These valleys, strips of fertility carved into an otherwise arid desert, provided the foundation for their cities. Each valley hosted its own political center, supported by irrigation canals that carried river water across fields of maize, beans, squash, and cotton. These engineered systems required constant maintenance and reflected strong local leadership and coordinated community labor.

Cities grew near river outlets, where agriculture, fishing, and trade converged. The Moche did not create a unified empire but a constellation of regional polities connected by shared artistic traditions, religious practices, and elite culture. Their urban centers included monumental temples, palaces, and courts made from millions of adobe bricks. These structures rose above the desert like sculpted mountains, decorated with murals that recorded scenes of warfare, rituals, and cosmology.

## Warriors, Rituals, and Moche Society

The Moche world was shaped by a warrior elite whose authority rested on both political power and religious roles. Ceramic vessels portray warriors in elaborate costumes and carrying clubs, shields, and spears. Some scenes depict ritual combat that may have served as a means to capture prisoners for ceremonial purposes.

Religion and political authority were deeply intertwined. Priests and rulers participated in complex ceremonies involving offerings, feasting, and the symbolic enactment of mythic narratives. One of the most striking discoveries related to Moche ritual life came from Huaca de la Luna, an enormous adobe temple near modern-day Trujillo.

## Huaca de la Luna and the World of the Temples

Huaca de la Luna, paired with the nearby Huaca del Sol, formed the spiritual and political heart of the southern Moche region. These stepped platforms were constructed in layers over centuries, each phase burying earlier temples beneath new ones. The walls of Huaca de la Luna feature vibrant murals depicting gods, mythical creatures, warriors, and ceremonial processions.

The most prominent figure is the so-called Decapitator, a deity shown with fanged teeth, clawed hands, and a crescent headdress. This god appears clutching a severed head and symbolizes the cycle of sacrifice and renewal that defined Moche religious ideology. The murals and offerings found within the temple suggest that rituals involving human sacrifice played a central role in maintaining cosmic order.

Archaeologists uncovered plazas stained with evidence of repeated ceremonial use. In some areas, remains of sacrificed individuals were found, often young men who may have been captured in battle. Their placement within the temple complex underscores the link between warfare, ritual, and elite power.

## Metalwork, Pottery, and Artistic Expression

The Moche excelled in metalworking and produced some of the finest gold, silver, and copper objects in pre-Columbian South America. Artisans hammered metals into delicate headdresses, nose ornaments, jewelry, and ceremonial vessels. Many pieces combined metals to create shimmering effects that reflect a sophisticated understanding of metallurgy.

Moche pottery is equally extraordinary. Their stirrup-spout vessels illustrate scenes of daily life, mythology, agriculture, medicine, and other elements of society with striking realism. Portrait vessels, ceramic heads modeled with lifelike detail, depict individuals with distinctive facial features or expressions. These vessels may represent specific people, including rulers, priests, or ancestors.

This artistic richness provides a vivid window into Moche society and shows everything from ceremonial finery to fishing, weaving, and farming.

## Climate, Catastrophe, and El Niño

For centuries, the Moche navigated the challenges of living between desert and sea. Their irrigation systems allowed them to flourish, but the environment was inherently unstable. One of the most influential forces in their history was the recurring climate pattern known as the El Niño–Southern Oscillation.

El Niño events brought powerful rains that flooded rivers normally reduced to narrow streams.

Floodwaters destroyed canals, toppled adobe structures, and buried farmland beneath layers of sediment. After intense flooding, prolonged droughts often followed and caused rivers to recede and crops to fail.

Archaeological evidence shows that the Moche experienced several major El Niño cycles. At Huaca de la Luna, layers of flood deposits and collapsing walls indicate repeated episodes of destruction. In some valleys, entire settlements were abandoned after catastrophic storms. The Moche rebuilt multiple times, but each cycle placed increasing strain on their political and social systems.

## Turmoil and Fragmentation

By the sixth and seventh centuries, signs of internal turmoil began to appear. Elite burials show changes in political hierarchy. Iconography shifts from orderly ritual scenes to images that suggest conflict and instability. Large fortified settlements emerged in some regions and indicate increased warfare or fear of attack.

Excavations at Huaca Cao Viejo, part of the El Brujo complex, revealed mass burial sites that speak to political upheaval. At the nearby tomb of the Señora de Cao, researchers uncovered the remains of a powerful female leader buried with elaborate metal ornaments. Her discovery challenged earlier assumptions that only men held elite roles and revealed a more complex political landscape.

Other settlements, such as the northern Moche capital at the Huacas del Sol y de la Luna, show evidence of violent endings. Burned structures, scattered remains, and rapid abandonment suggest internal conflict or invasion by rival groups.

By about 800 CE, the Moche world fractured. Some valleys were depopulated, while others saw new political systems emerge. The shift was not a simple collapse but a long process of transformation driven by environmental instability, social change, and competition among regional powers.

## Rediscovery and Modern Insights

The eroded temples and weathered murals of the Moche attracted scholarly attention in the nineteenth century, but systematic excavation began in earnest in the late twentieth century. Archaeologists such as Rafael Larco Hoyle helped establish the first ceramic typologies and enabled researchers to reconstruct Moche chronology and regional variations.

In recent decades, major discoveries have reshaped understanding of Moche leadership and ritual practices. The tomb of the Lord of Sipán, uncovered in 1987 in the Lambayeque Valley, revealed an elite burial unmatched in wealth and preservation. The tomb contained gold ornaments, jewelry, ceramics, and the remains of sacrificial attendants. Its contents illuminated the role of warrior-priests and the ceremonial power wielded by Moche leaders.

Scientific studies now combine climate data, isotope analysis, and high-resolution mapping to explore how environmental shifts influenced settlement patterns. Canal systems have been traced using satellite imagery. Flood layers have been matched to known El Niño events. These tools allow researchers to build a more precise timeline of Moche resilience and vulnerability.

## A Legacy of Adaptation and Innovation

The Moche did not vanish. Their descendants continued to adapt and formed new cultural groups and political systems that thrived in later centuries. Their artistic and engineering achievements influenced subsequent cultures along Peru's northern coast, including the Lambayeque, Chimú, and later the Inca.

Their irrigation knowledge continued to shape agriculture in the region, and many canals follow ancient pathways still recognizable today. The Moche demonstrated how a civilization could harness an unforgiving environment through innovation, community labor, and ritual dedication.

As storm clouds gather over Peru's northern desert, the eroded pyramids of the Moche seem fragile against a backdrop of shifting weather and silent valleys. Their world was built on rivers that offered both abundance and risk. Their cities flourished through ingenuity, artistry, and carefully maintained irrigation. Yet they faced forces beyond human control, including climate cycles, floods, droughts, and social tensions that challenged the foundations of their society.

Today, the Moche story is reconstructed from adobe walls, burial chambers, murals, and the landscape itself. Their sudden silence reminds modern readers of how environmental change can disrupt even well-established civilizations. The remains of their river-fed cities stand as a testament to resilience and creativity and echo across a landscape that once supported one of the most remarkable cultures of ancient South America.

# Chapter 8
# Cities of the Jaguar Lords – Palenque, Yaxchilán, and Maya Power on the Rivers

Dawn rises slowly over a jungle river. Mist curls above the still water and drifts toward dense walls of greenery that conceal ancient stone. As the light strengthens, carved stelae appear along the riverbank, their glyphs catching the soft glow of morning. The stones speak of kings whose lives were etched into history, alliances made, battles won, and betrayals remembered. These are the cities of the Jaguar Lords, the riverine Maya capitals whose power flowed with the currents that sustained them.

## Kingdoms Along the Water

Rivers defined much of Maya life. In the humid lowlands of Chiapas and Guatemala, waterways served as arteries of commerce, communication, and political influence. Settlements rose along these courses, carving footholds into steep hills or clearing open spaces for monumental centers. Among the most powerful of these river-based kingdoms were Palenque on the Usumacinta's tributaries and Yaxchilán along the river itself.

Palenque, surrounded by lush mountains and fed by abundant springs, became a jewel of Maya architecture. Its palaces, aqueducts, and temples integrated natural water features into ceremonial life. Inscriptions reveal a dynasty shaped by triumph and hardship, leading to one of the most celebrated reigns in Maya history, that of K'inich Janaab' Pakal, known as Pakal the Great.

Yaxchilán, perched on a bend of the Usumacinta, held command over key river passages. Its rulers styled themselves as Jaguar Lords and projected power through monumental architecture and a fierce visual identity carved into stelae and lintels. The river below served as both defensive barrier and supply route and linked Yaxchilán to allies, tributaries, and rivals.

These kingdoms, and others like them, created networks of influence that shaped Classic Maya politics for centuries.

## Dynastic Politics and Royal Narratives

Maya kingship was theatrical, strategic, and rooted in mythology. Inscriptions carved into stone monuments recorded births, deaths, wars, marriages, accessions, and rituals. The Maya wrote history not as a neutral chronicle but as a demonstration of lineage and divine favor.

At Palenque, dynastic texts revealed a long line of rulers stretching back centuries. For many years, scholars struggled to decipher the glyphs. Early theories dismissed Maya writing as symbolic, but breakthroughs in the mid-twentieth century, led by researchers such as Tatiana Proskouriakoff and Yuri Knorosov, proved the script recorded historical events.

With decipherment, Palenque's rulers came into sharp focus. Pakal inherited the throne in 615 CE and ruled for nearly seven decades. Under his leadership, Palenque reached new heights of architectural innovation. His successors continued his legacy, commemorating their lineage in the Temple of the Inscriptions, where Pakal's tomb was discovered deep within a steep pyramidal structure.

At Yaxchilán, inscriptions portray rulers such as Shield Jaguar II and Bird Jaguar IV, whose reigns were marked by warfare, alliances, and ceremonial performances. Many carved lintels show queens participating in bloodletting rituals and drawing cords through their tongues to invoke visions, acts that reinforced the supernatural legitimacy of royal families.

These written records reveal a world of shifting loyalties. Marriage alliances forged political bonds between distant cities. Captives taken in battle became symbols of dominance. Ritual performances validated authority before the gods and the people.

## Architectural Innovation and Environmental Mastery

Riverine Maya cities exploited their landscapes with precision. Palenque, with its plentiful springs, built elaborate aqueducts to manage water flow. Some structures channeled water beneath palaces, creating controlled streams that cooled interior spaces and prevented flooding during the rainy season. The Palace complex, a maze of courtyards, galleries, and towers, exhibited refined engineering and artistic skill.

Yaxchilán's monuments relied on strategic placement. Buildings rose along limestone terraces with commanding views of the river. Stairways, plazas, and structures staged the movement of elites and visitors and guided them through ritual spaces shaped to reinforce royal presence.

Many Maya cities used the river not only as a resource but as a boundary. Yaxquilán's rival, Piedras Negras, sat upriver and competed for control of key routes. Warfare between river kingdoms often involved capturing tribute, installing sympathetic leaders in subordinate towns, and making strategic displays of power along riverbanks accessible to both allies and enemies.

## Discoveries That Transformed Maya Studies

Much of what is known today about Maya river kingdoms comes from breakthroughs made in the last century. Early explorers mapped temples hidden beneath dense vegetation, but the full meaning of Maya art and inscriptions remained elusive until the decipherment of the script.

Tatiana Proskouriakoff's work in the 1960s demonstrated that stelae recorded birth and accession events and proved that Maya inscriptions chronicled real history. Yuri Knorosov's discovery that Maya writing was phonetically based provided the foundation for modern decipherment. Together, these breakthroughs transformed Palenque, Yaxchilán, and many other sites from mysterious ruins into readable archives.

Inscriptions at Yaxchilán revealed intricate political relationships, including alliances with nearby centers such as Bonampak, famous for its detailed murals depicting warfare and ritual. These texts also document rivalries that shaped the region, including conflict with Piedras Negras and shifting alliances with smaller polities along the Usumacinta.

At Palenque, decipherment brought clarity to its architectural program. Each temple and palace building found new meaning as scholars linked inscriptions to construction phases, historical events, and ritual purposes.

## Warfare and Resource Pressure

Maya kingship was inseparable from warfare. Capturing enemies, especially elite captives, validated a ruler's strength and ensured divine favor. River cities such as Yaxchilán and Piedras Negras frequently recorded such events and often showed the ritual of presenting captives before the king.

However, warfare placed heavy demands on populations. Resources used to support military campaigns strained local economies. The need to maintain alliances created political vulnerabilities, and losing a war could destabilize entire regions.

Environmental pressure compounded these challenges. Deforestation around rivers led to soil depletion and erosion. As populations grew, the demand for fields, firewood, and building materials increased. Climate studies indicate that severe droughts struck many parts of the Maya lowlands during the eighth and ninth centuries. Rivers shrank. Crops failed. Populations suffered.

The combination of warfare, ecological strain, and political instability contributed to the decline of many river-based kingdoms. Yaxchilán's monuments cease after the late eighth century. Palenque's inscriptions end even earlier, with its last known date in 799 CE. Piedras Negras collapsed soon after, leaving its plazas abandoned to the jungle.

## Shifting Landscapes and the End of Maya Power on the Rivers

The fall of these cities was not sudden but part of a broader regional process known as the Classic Maya Collapse. As drought cycles intensified, resources dwindled, and political institutions weakened, populations began to disperse. Some moved to more stable regions. Others reorganized into smaller communities less dependent on large ceremonial centers.

River kingdoms, once among the most powerful, were especially vulnerable. Their political structures relied on continuous access to water, fertile floodplains, and control over river routes. When these factors faltered, so too did the foundations of their authority.

Yet the Maya did not disappear. Their descendants continued to live throughout the region and adapted to new political, economic, and environmental realities. Traditions persisted and evolved, preserving cultural continuity even as the great cities fell into silence.

As the morning mist lifts from the jungle river, carved stones glimmer in the sunlight, their glyphs offering glimpses into a world shaped by power, ambition, and spiritual depth. Palenque, Yaxchilán, Piedras Negras, and other river kingdoms once commanded vast networks of alliances and influence. Their rulers recorded triumphs and tragedies in stone and left behind stories that modern scholars are only now able to fully read.

Their decline reflects the vulnerability of complex societies facing environmental stress, political fragmentation, and shifting resource landscapes. Yet their legacy endures through inscriptions, architecture, oral traditions, and the living Maya communities who preserve cultural knowledge today.

The river flows onward, carrying echoes of the Jaguar Lords who once shaped their world from its banks. Their cities, now reclaimed by forest, remain as monuments to human ingenuity, resilience, and the ever-changing balance between society and the environment.

# Chapter 9
# Forgotten in the Foothills – The Hohokam Canals and the Vanished Cityscape of Phoenix

A drone glides above the sprawling grid of modern Phoenix, its freeways, neighborhoods, and golf courses stretching toward distant mountains. Beneath this built landscape, beneath parking lots, subdivisions, and multilane roads, lie faint traces of canals dug more than a thousand years ago. The desert sun blazes across the Salt River Valley and reminds viewers how improbable a major metropolis once seemed in this arid region. Yet long before Phoenix rose from the desert floor, the Hohokam engineered one of the most extensive irrigation networks in ancient North America and turned inhospitable land into a thriving agricultural world.

## Cities Shaped by Water

The Hohokam flourished in central and southern Arizona from about 300 to 1450 CE. Their homeland lay in one of the driest environments on the continent, where summer temperatures soared and annual rainfall remained scarce. To survive here required more than simple adaptation. It required reimagining the desert itself.

The Salt and Gila Rivers, fed by seasonal snowmelt from distant mountains, offered the region's most reliable source of water. The Hohokam recognized their potential and transformed them through engineering. Over centuries, they built an enormous system of canals, some stretching more than twenty miles, that diverted river water into hundreds of miles of branch channels. These canals irrigated thousands of acres, supported large communities, and enabled surplus crop production.

Their agricultural fields produced maize, beans, squash, cotton, agave, and other crops suited to desert farming. Cotton became particularly valuable, both for weaving and trade. With food security established, population centers grew along canal routes and formed a network of villages and ceremonial complexes throughout the valley.

## Engineering on a Monumental Scale

Hohokam canals were feats of planning, calculation, and community labor. Engineers had to determine where to divert river water, how steep the canal slopes should be, and how to maintain flow without causing erosion or flooding. Some main canals measured more than sixty feet wide and fifteen feet deep and required excavation by hand with stone and wooden tools.

The canal system followed the contours of the land with remarkable precision. Gentle gradients allowed gravity to move water across long distances while preventing blockages. Communal labor crews maintained the canals seasonally, clearing debris and repairing breaches after floods.

Archaeologists have traced multiple phases of canal expansion and show that Hohokam engineers improved and extended the system over centuries. By 1200 CE, the Salt River Valley supported one of the largest premodern irrigation networks in the world.

## Platform Mounds and Ballcourts

Agriculture fueled the growth of Hohokam cultural centers that featured platform mounds, ballcourts, and villages arranged around plazas. These architectural forms reflect a well-organized society with ceremonial and political complexity.

Platform mounds, flat-topped earthen structures, served as foundations for buildings used in elite residences, rituals, or administrative activities. The largest, such as those at Pueblo Grande and La Ciudad, required vast amounts of soil transported in baskets. Multiple construction phases show long-term planning and periodic renewal.

Ballcourts, shaped like elongated ovals with sloping sides, indicate strong cultural ties with Mesoamerica. The Hohokam participated in a ballgame tradition shared across regions and incorporated ritual and social functions. Ceremonies, gatherings, and trade likely centered around these courts.

Surrounding villages of pit houses and later above-ground structures suggest a stable population with specialized crafts. Pottery painted with red-on-buff designs, shell ornaments traded from the Gulf of California, and stone tools reflect both local production and long-distance exchange.

## A Complex Society in the Desert

The Hohokam world was not an isolated one. They interacted with neighboring cultural groups across the Southwest and participated in extensive trade networks. Macaws, copper bells, obsidian, and shell goods connect the Hohokam to societies in Mesoamerica and beyond.

Their agricultural abundance supported dense populations estimated in the tens of thousands. Villages aligned along the canals created a cultural landscape tied together by shared rituals, cooperative water management, and regional gatherings. Seasonal festivals, craft production, and community decision-making may have taken place at large ceremonial centers.

## Theories of Abandonment

By the 1400s, many Hohokam settlements were abandoned, and the canal system fell into disuse. The reasons remain debated, and archaeologists increasingly view the process as gradual rather than sudden. Several interconnected factors likely contributed.

Drought cycles played a significant role. Paleoclimate studies show that prolonged droughts affected the Southwest during the Medieval Climate Anomaly. Reduced river flow would have made canal maintenance and irrigation more challenging. Crop failures may have weakened communities already stretched thin by environmental stress.

Flooding presented the opposite but equally devastating problem. Major floods periodically damaged canals, washed away walls, filled channels with sediment, or shifted river courses entirely. Rebuilding required enormous labor and coordination. Repeated flood events may have overwhelmed the system.

Social stress emerged as populations grew. Larger communities required more irrigation water, which increased competition and the need for cooperation. Disputes over water rights, leadership changes, or internal conflict may have destabilized village networks.

External conflict is also considered. Some areas show evidence of fortified positions or burned structures that suggest violence. As resources dwindled, tensions with neighboring groups may have escalated.

By the late 1200s to 1400s, many residents migrated north or east and joined communities that evolved into the O'odham, Hopi, and other Indigenous groups of the region. Their descendants maintain oral histories that describe migrations from the river valleys and emphasize environmental change, spiritual guidance, and community adaptation.

## Rediscovery and the Work of Archaeology

Modern recognition of the Hohokam began in the late nineteenth century, but systematic excavation and mapping in the twentieth century transformed understanding of their achievements. Archaeologists identified canal segments beneath modern Phoenix and revealed their extraordinary scale. Excavations at Pueblo Grande, Mesa Grande, Snaketown, and La Ciudad uncovered platform mounds, ballcourts, pit houses, and artifacts that illustrated a sophisticated agricultural society.

Dendrochronology and sediment analysis allowed researchers to link canal damage and abandonment to specific drought and flood events. Radiocarbon dating refined the chronology of village development. Remote sensing and aerial photography helped trace canal routes buried under modern infrastructure.

The realization that Phoenix was built atop hundreds of miles of ancient irrigation channels reshaped public appreciation for the region's deep human history. Many modern canals operated by the Salt River Project follow or parallel ancient Hohokam alignments and demonstrate the enduring practicality of their engineering.

## A Legacy Beneath the Modern City

Today, the Salt River Valley remains a living testament to Hohokam ingenuity. Though their cities vanished centuries ago, their canal lines continue to guide urban development. Visitors can walk through preserved sections at Pueblo Grande Museum or view reenvisioned watercourses integrated into modern parks. Engineers and hydrologists often note how closely the modern system mirrors ancient designs.

For descendant communities such as the Akimel O'odham and Tohono O'odham, the Hohokam are part of a larger ancestral story. Oral traditions describe migrations, cultural transformations, and spiritual connections to the rivers that sustained their predecessors. Archaeology adds layers of understanding and reveals how these ancient engineers reshaped the desert through collective effort and innovation.

As the drone pulls back from modern Phoenix, the outline of the valley's ancient canals seems to rise from the shadows, ghost lines beneath pavement, fields, and neighborhoods. The Hohokam carved a civilization from the desert through mastery of water and built communities that flourished for more than a millennium. Their decline reflects the challenges of managing fragile environments under changing climate conditions, growing populations, and shifting social dynamics.

Yet their legacy endures, not only in the traces of their irrigation network but in the living cultures connected to them and in the modern city shaped in part by their achievements. The story of the Hohokam reminds us that even in the harshest environments, human ingenuity can transform a landscape and sustain a civilization, though always at the mercy of forces greater than the canals that once carried life across the desert floor.

# Chapter 10
# Storm Over the Highlands – The Collapse of the Wari Heartland

Thick Andean clouds roll across terraced mountains and drift low over stone walls and open plazas. Below the mist lie the remains of administrative compounds once alive with movement, messengers crossing courtyards, storage rooms filled with maize and potatoes, and officials overseeing a network of roads stretching far beyond the horizon. Today these highland complexes sit silent, carved into slopes where wind now moves through doorways meant for a very different age. This is the heartland of the Wari Empire, an early Andean power whose rise shaped centuries of South American history, and whose fall unfolded beneath gathering storms of drought and political unraveling.

## Building an Empire in the Highlands

The Wari civilization emerged around 600 CE in the Ayacucho region of Peru. Over the next several centuries, it expanded into one of the first major empires of the Andes and influenced the later Inca. Its capital, near the modern city of Ayacucho, covered several square miles and included monumental architecture, gridded neighborhoods, vast storage facilities, and administrative centers connected by a planned network of roads.

Wari architects developed a distinctive style. Rectilinear compounds enclosed by high walls were organized around central patios. These complexes served as administrative hubs and housed officials, craftspeople, and storerooms. Their design appears throughout the empire, from the southern site of Pikillaqta to the northern centers of Huari and Viracochapampa. This consistent architecture suggests strong central planning and a desire to project authority across diverse regions.

## Roads, Storage, and Imperial Order

The Wari developed an extensive roadway system long before the Inca expanded and refined it. These roads linked highland administrative centers with coastal valleys and remote mountain settlements. Pack llamas carried goods along these routes and moved textiles, pottery, maize, coca leaves, obsidian, and other resources crucial for administration and ritual.

Storage was essential to the Wari economic system. Hundreds of rectangular storehouses, known as collcas, were built to collect and distribute food, tools, and textiles. These facilities helped manage agricultural surplus and buffered communities against environmental instability. By controlling storage and distribution, Wari elites strengthened their political authority and secured loyalty from local populations.

Craft workshops within Wari compounds produced fine ceramics, textiles, and metalwork that carried symbolic imperial imagery. These goods circulated throughout the empire and reinforced cultural unity and elite identity.

## Planned Cities and Social Transformation

Wari urban centers were more than administrative hubs. They reshaped surrounding landscapes through terracing, irrigation, and controlled settlement planning. Pikillaqta, near Cusco, exemplifies Wari urban design, a gridded city with long, narrow corridors, high walls, and precisely aligned streets. It likely served as both a ceremonial site and a regional administrative center.

Other centers, such as Conchopata, played key roles in ritual and artistic production. Ceramics found there depict deities later incorporated into Inca iconography and suggest shared religious foundations across centuries.

The Wari also implemented agricultural innovations, including terracing and water management strategies, that stabilized food production in challenging environments. Their influence spread widely and transformed social and political life throughout the central Andes.

## Evidence of Decline

For centuries, the Wari system thrived through centralized planning, agricultural surplus, and political integration. Yet by about 1000 CE, many Wari centers show signs of abandonment or reduced occupation. The empire that once unified much of the Andes fragmented into regional polities.

Recent archaeological discoveries point to a combination of environmental and political factors behind the decline.

Drought appears to have played a major role. Paleoclimate studies, including lake sediment analysis, ice core data, and isotopic research, show that a series of severe droughts struck the central Andes during the late first millennium. These droughts reduced crop yields and dried streams essential for terraced agriculture. With less water available, food surplus collapsed and undermined the foundation of Wari administrative power.

Decentralization followed environmental stress. As agricultural production faltered, regional elites may have asserted greater autonomy. Wari compounds were abandoned or repurposed. The once-unified road and storage network began to break down.

Population movement intensified as as communities left drought-stricken valleys for areas with more reliable water sources. Some groups fortified settlements and suggest increased conflict or competition over limited resources.

Archaeologists see little evidence of widespread violent destruction at Wari sites. This indicates the collapse was gradual rather than catastrophic. Administrative centers dwindled and political integration dissolved from within.

## The Shadow of Influence: Wari and the Inca

Though the Wari Empire fell centuries before the rise of the Inca, its legacy echoed throughout the Andean world. The Inca drew heavily from Wari models and adapted their architecture, governance, and infrastructure.

Inca roadways expanded earlier Wari routes and linked highland and coastal territories in an integrated network. Many Inca roads follow Wari alignments and demonstrate the durability of the earlier system.

Inca administration adopted strategies familiar from Wari centers, storage complexes, state-sponsored craft production, and the organization of labor through hierarchical structures. Even the concept of a planned imperial capital, seen later at Cusco, may have roots in Wari urban planning.

Religious imagery also shows continuity. Deities portrayed on Wari ceramics, textiles, and pottery, such as the Staff God, reappear in Inca iconography and reflect shared spiritual traditions across Andean cultures.

The Inca likely viewed Wari ruins as ancestral and even sacred. They built upon Wari foundations and incorporated earlier architectural forms into their own imperial vision.

## Archaeology and Rediscovery

The Wari heartland remained little known outside local communities until the twentieth century. Early surveys identified large ruins but misunderstood their significance. It was not until later excavations led by Peruvian and international archaeologists that the scale of the Wari Empire became clear.

Excavations at the Wari capital uncovered administrative compounds, residential zones, elite tombs, and evidence of large-scale craft production. At Pikillaqta, researchers mapped the city's grid layout and identified features that linked it to highland imperial planning. Recent work at sites such as Espíritu Pampa has revealed continued influence in remote regions.

New technologies, including drone mapping, ground-penetrating radar, and soil analysis, are helping archaeologists reconstruct Wari political organization, climate impacts, and demographic changes with greater precision.

As clouds sweep across the Andean highlands, the silent terraces and abandoned compounds of the Wari heartland bear witness to an empire that rose through innovation and organization yet faltered under the pressure of environmental change and political fragmentation. The Wari transformed landscapes with roads, terraces, and gridded cities and crafted systems later refined by the Inca.

Their collapse was not dramatic but gradual and spread across generations as drought strained resources and decentralized leadership eroded imperial cohesion. What remains are stone foundations, scattered ceramics, and the faint outlines of a civilization that shaped Andean history in ways still being uncovered.

The storm passes over the terraced mountains and leaves the Wari heartland in quiet shadow. Its story reminds modern readers that even the most well-planned societies depend on the stability of their environment and the resilience of their political systems. The Wari navigated both with skill for centuries and left behind a legacy that continued to shape the Andes long after their own cities grew silent.

# Chapter 11
# The Red City in the Dunes – Norte Chico's Monument Builders

Ancient mounds rise from a stark Peruvian coastal desert, their silhouettes softened by centuries of wind yet unmistakably monumental. At their feet, sunken circular plazas open toward the sky, carved into sandy ground that stretches toward distant waves crashing against the Pacific shore. The landscape feels both barren and timeless. Dry riverbeds cut across the plain and hint at seasonal flows that once sustained an astonishingly early experiment in urban life. This is the world of the Norte Chico civilization, also known as Caral, where some of the earliest cities in the Americas took shape almost five thousand years ago.

## Cities Older Than Pyramids

Long before the rise of the Inca, centuries before the Moche or Wari, and even predating most ancient civilizations in the Old World, the people of Norte Chico built cities along four river valleys north of Lima. Between roughly 3000 and 1800 BCE, these communities created monumental architecture that challenged long-held assumptions about early urbanism in the Andes.

The largest and best studied center, Caral, sits in the Supe Valley. Massive platform mounds, often called pyramids, stand beside wide plazas and ceremonial spaces connected by stairways and terraces. These structures, built of stone and earth rather than carved masonry, reveal sophisticated planning. Caral alone contains more than 30 major architectural constructions, including the Great Pyramid, which rises nearly 100 feet.

Other urban centers, such as Aspero on the coast and inland sites such as Bandurria, Vichama, and Miraya, formed part of a broader network of settlements spread across river valleys and desert terraces. Together, they create one of the oldest known urban landscapes anywhere in the Americas.

## Irrigation, River Management, and Coastal Abundance

Norte Chico cities developed in a coastal region where rainfall is minimal. The people who built them relied heavily on river-fed irrigation. Early engineers diverted river water into canals that nourished fields of cotton, gourds, squash, beans, and other crops suited to the desert environment.

Cotton played a central economic role. Unlike many early civilizations that relied on grain surpluses, Norte

Chico communities used cotton to make fishing nets, textiles, bags, and ceremonial items. Cotton-based technology linked coastal and inland settlements into a shared economy. Inland farmers produced cotton, and coastal groups used nets woven from that cotton to harvest anchovies, sardines, and other marine resources.

This river-to-coast relationship created a stable food system and supported population growth. The abundance of fish protein allowed inland cities to flourish even when agriculture alone might not have sustained them. As a result, monumental architecture rose in a region often considered too dry for large civilizations.

## Monumental Platforms and Sunken Plazas

Architecture defines Norte Chico's most striking achievements. Platform mounds vary in size but share core characteristics, layered construction, stone retaining walls, and central staircases. Many were rebuilt multiple times, each phase adding new walls or floors, which suggests generations of community labor.

Equally impressive are the sunken circular plazas, architectural elements that appear repeatedly across Norte Chico sites. These sunken courts may have served for ceremonies, music performances, or gatherings tied to seasonal and ritual cycles. They created a spatial focus that drew people inward and emphasized communal participation.

Archaeologists have found evidence of burned offerings, reed baskets used in construction, and musical instruments such as bone flutes. These objects help reconstruct a cultural life that valued ceremony, shared identity, and coordinated labor rather than warfare or conquest.

## A Society Built on Cooperation

One of the most remarkable aspects of Norte Chico is the apparent lack of warfare or defensive structures. Unlike many other early civilizations, Norte Chico left little evidence of fortifications, weaponry, or violent conflict. Instead, its social and political systems appear to have been built on cooperation and collective labor.

Communal construction projects, including the building and rebuilding of platform mounds, required organizing workers across multiple settlements. This suggests a leadership structure capable of mobilizing large groups but not necessarily one based on military power. Elite residences are modest, and burial evidence is sparse, indicating a society that may have expressed status and authority differently than later Andean cultures.

Feasting rituals, musical performances, and ceremonies conducted in sunken plazas likely helped maintain social cohesion. Textile production, irrigation maintenance, and trade networks further reinforced connections between communities.

## Peaceful Longevity and Gradual Decline

For more than a thousand years, Norte Chico cities thrived. Their longevity, from roughly 3000 to 1800 BCE, makes them some of the most enduring early urban settlements in human history.

Their decline appears to have been gradual. Environmental shifts, including changes in river courses and periodic droughts, likely strained agricultural systems. Coastal communities faced fluctuations in marine resources tied to climate cycles such as El Niño events.

As rivers deposited new sediment, earlier canals may have become unusable and prompted relocation or abandonment. Over time, people dispersed into smaller communities or migrated into new regions and carried with them technologies and social traditions that influenced later Andean societies.

By about 1800 BCE, large ceremonial centers such as Caral saw declining construction activity and many sites were partially abandoned.

## How Rediscovery Changed the Andean Timeline

The modern rediscovery of Caral and the Norte Chico civilization reshaped archaeological understanding of Andean history. Although explorers documented some ruins in the late nineteenth century, the true significance of Caral remained unclear until the 1990s, when Peruvian archaeologist Ruth Shady Solís led comprehensive excavations.

Radiocarbon dating revealed astonishing results. The great mounds and plazas of Caral were far older than previously believed, which made the Supe Valley one of the earliest cradles of civilization in the world. It was contemporary with Egypt's Old Kingdom and Sumer's early cities.

This revelation forced scholars to reconsider assumptions about how and where complex societies first developed. It showed that monumental architecture, planned cities, irrigation systems, and long-distance economic networks emerged in the Andes more than fifteen hundred years earlier than once thought.

Norte Chico also challenged the idea that early civilizations must rely on warfare, grain agriculture, or monumental stonework. Instead, it demonstrated that cooperation, cotton-based trade, and a marine-supported economy could produce urban complexity of remarkable scale.

## Influence on Later Andean Cultures

Although separated by centuries, Norte Chico's legacy can be traced in architectural and cultural traditions that appear throughout Andean history. The emphasis on communal labor, formal plazas, and platform architecture reappears in later civilizations such as the Chavín, Moche, Wari, and Inca.

Certain symbols and ceremonial practices may have roots in the early Supe Valley. The Andean concept of organizing cities around ritual centers, supported by agricultural surplus and coordinated labor, echoes the Norte Chico model. Even the idea of powerful but nonmilitaristic leadership, evident in some later Andean polities, finds precedent in Caral.

As waves crash against the distant coast and winds move across the desert dunes, the platform mounds and sunken plazas of Norte Chico stand as reminders of one of the world's earliest urban experiments. Their builders created a civilization shaped not by conquest but by irrigation, cooperation, and shared ceremonial life. Their cities rose in a landscape where survival demanded innovation, patience, and unity.

The rediscovery of Norte Chico reframed the timeline of Andean civilization. It revealed that complex societies emerged here millennia earlier than once imagined. Their decline was gentle, their influence enduring, and their story a powerful testament to humanity's ability to build monumental worlds even in the most unlikely environments.

The red city in the dunes may be quiet now, but the echoes of its builders continue to reshape how the world understands the deep past of the Andes.

# Chapter 12
# The Tunnels Beneath the Temple — Chavín and the Roots of Andean Civilization

Dark stone corridors twist beneath a highland temple, their surfaces worn smooth by centuries of ritual passage. Thin shafts of sunlight filter through narrow ventilation openings and illuminate carved fangs, serpent motifs, and enigmatic faces that seem to emerge from the walls. The air is cool and still, carrying faint echoes of water flowing through unseen channels. This is the underground world of Chavín de Huántar, one of the earliest and most influential ceremonial centers in the Andes, a place where architecture, religion, and engineering shaped a spiritual landscape that endured long after its abandonment.

## A Highland Crossroads of Ceremony and Power

Located at a strategic convergence of mountain valleys in central Peru, Chavín de Huántar flourished between roughly 900 and 200 BCE. Its position connected the Pacific coast, highland plateaus, and Amazonian foothills, which made it an ideal crossroads for trade, pilgrimage, and cultural exchange. Over centuries, Chavín became a major ceremonial hub whose influence spread across the central Andes.

The site's monumental architecture, including platform temples, sunken plazas, and stone stairways, formed a ritual complex that guided visitors through increasingly restricted spaces. Pilgrims entered wide plazas for public ceremonies before moving into narrower passageways that led toward the heart of the temple. Here, hidden within the labyrinth, stood the most iconic object of Chavín religion: the Lanzón, a carved granite monolith rising more than fifteen feet.

The Lanzón depicts a fanged supernatural being with flowing hair and crossed fangs, part feline, part serpent, and part human. Its exact meaning remains debated, but scholars agree it represented a powerful deity whose presence dominated Chavín's ceremonial core.

## Subterranean Galleries and Ritual Architecture

Chavín's architecture is notable not only for its monumental scale but also for its complex system of subterranean galleries. These stone-lined tunnels run beneath the temple platform and form a labyrinth designed to confuse and disorient. Some galleries are pitch black. Others contain shafts of light that create intentional visual effects during ritual performances.

Many passages narrow unexpectedly or make sudden turns, forcing participants to navigate through sound, touch, and memory rather than sight. Acoustic engineering allowed priests to speak or blow conch-shell trumpets in one chamber while the sound reverberated through distant corridors, creating the impression of voices from the earth or the gods themselves.

The galleries vary in size and orientation, which suggests different ceremonial functions. Some may have been used for processions, trance rituals, or initiation events. Others held niches for offerings, altars, or ritual paraphernalia. The overall effect was to create a sensory environment in which architectural design shaped spiritual experience.

## Religious Iconography and Symbolic Power

Chavín art helped unify the diverse groups who traveled to the site. Its iconography includes fanged deities, jaguars, condors, snakes, and hybrid beings that combine elements of rainforest predators, highland animals, and mythical creatures. These images appeared on stone stelae, lintels, pottery, textiles, and carved bone.

One of the most famous pieces is the Tello Obelisk, a four-sided stone carving that depicts intertwined creatures in a dense symbolic narrative. Its imagery may reference agricultural fertility, duality, or mythic origin stories. Another key object, the Raimondi Stele, portrays a staff-bearing deity whose image shifts depending on how the stone is viewed, an example of Chavín artists' mastery of optical illusion and layered symbolism.

This shared visual language traveled widely and influenced ceramics, textiles, and architectural styles across the Andes. Communities hundreds of miles away adopted Chavín iconography, which suggests that its religious system offered spiritual legitimacy and cultural identity to distant groups.

## Water, Engineering, and Environmental Mastery

Chavín's builders integrated water into the ceremonial experience. The temple complex contains channels that redirect river water through specific locations and produce sounds, vibrations, or controlled flooding that may have been part of ritual events. These channels also served practical purposes by preventing erosion and stabilizing the temple's foundations during rainy seasons.

The site's drainage system is remarkably sophisticated. Engineers carved stone conduits with precise gradients that allowed water to flow smoothly and protected subterranean chambers and surface structures from damage. The integration of environmental management with ceremonial architecture demonstrates the technical expertise of Chavín's designers.

## A Pilgrimage Center Without Fortifications

Unlike later Andean states, Chavín appears to have been a ceremonial and religious center rather than a militarized capital. The absence of defensive walls, weapons caches, or evidence of warfare indicates that its power rested on spiritual authority rather than military force.

Pilgrims likely traveled from distant regions to participate in rituals, consult priests, or receive blessings tied to agricultural fertility, healing, or political legitimacy. The broad spread of Chavín-style art suggests that visiting groups adopted the site's religious forms and carried them back to their home communities.

This diffusion helped shape a shared Andean cultural sphere long before the rise of large states such as the Wari or Inca.

## Decline and Abandonment

Chavín's decline around 200 BCE appears to have been gradual rather than sudden. Environmental change, including landslides and shifts in river patterns, likely damaged key parts of the temple complex. The site sits near steep mountain slopes, which made it vulnerable to natural disasters. Evidence of flooding and debris flow suggests that environmental instability may have undermined Chavín's infrastructure and ritual functions.

As other regional centers emerged, the flow of pilgrims diminished. Political landscapes shifted, and

Chavín's once dominant religious influence waned. By the beginning of the first millennium CE, the site was largely abandoned, although its legacy endured in symbols and architectural concepts adopted by later cultures.

## Rediscovery Through Excavation and Science

Archaeological interest in Chavín began in the early twentieth century with the work of Julio C. Tello, considered the father of Peruvian archaeology. Tello recognized Chavín as a foundational culture in the Andean world and identified many of its most important artifacts.

Recent excavations have expanded this understanding. Archaeologists such as John Rick and his team have uncovered new galleries, plazas, and ritual features previously unknown. Advances in 3D mapping, underground imaging, and paleoenvironmental analysis have revealed that much of Chavín's architecture remains buried or unexplored, which suggests a far more extensive ceremonial complex than once believed.

These discoveries show that Chavín's subterranean world was more intricate than imagined, an engineered environment of light, sound, and symbolic meaning designed to shape spiritual experience.

## Legacy of the First Andean Religion

Chavín's influence on later Andean civilizations is profound. Elements of its iconography, architectural planning, and ritual strategies appear in the Wari, Moche, Nazca, and Inca cultures. The concept of a pilgrimage center, a hierarchy of ritual specialists, and the use of monumental art to project religious authority all trace roots to Chavín's innovations.

The Inca, centuries later, adopted similar ideas about sacred landscapes, ceremonial manipulation of water, and integration of architecture with cosmic symbolism. Although the builders of Chavín left no written records, their spiritual vision shaped Andean civilization for millennia.

In the dim corridors beneath the temple, carved fangs and serpentine figures still loom from the walls and remain as reminders of a time when architecture and belief were inseparable. Chavín de Huántar stands as one of the earliest great ceremonial centers of the Andes, a place where engineering, artistry, and religion converged into a powerful cultural force.

Its decline was gradual, but its impact endured. The tunnels, plazas, and monoliths continue to reveal how ancient peoples understood their world and manipulated space, sound, and symbol to shape religious experience. Chavín's story, hidden beneath stone and earth for centuries, remains a cornerstone for understanding the roots of Andean civilization and a testament to human imagination carved deep into the highland mountains.

# Chapter 13
# Islands in the Lake — The Purépecha Capitals That Defied the Aztec World

Morning light moves softly across the surface of Lake Pátzcuaro and turns the water into shifting silver. Small islands rise from the lake like quiet sentinels, their outlines mirrored in the stillness. Along some shorelines, low stone walls and foundations emerge from the earth, remnants of cities that once commanded this highland basin. Here, in the cool volcanic plateau of western Mexico, the Purépecha Empire built fortified capitals, metalworking centers, and island settlements that stood firm against the power of the Aztec world.

## A Highland Empire with Island Roots

The Purépecha, also known as the Tarascans, flourished in the region of Michoacán from roughly 1100 to 1530 CE. Their homeland, defined by volcanic peaks, fertile valleys, and a series of lakes, most prominently Lake Pátzcuaro, gave rise to a distinctive cultural and political system that rivaled the great empires of Mesoamerica.

Unlike many neighboring cultures, the Purépecha developed strong urban centers connected by roads, administrative districts, and ceremonial sites. The lake itself anchored their world. Island communities served as refuges, religious centers, and hubs of craft production. Settlements on the mainland ringed the water and formed a defensive and economic network that allowed the empire to thrive.

## Tzintzuntzan, Capital of the Purépecha World

The Purépecha capital, Tzintzuntzan, rose on the eastern shore of Lake Pátzcuaro. Its name means "Place of the Hummingbirds," yet the site conveyed strength and authority rather than delicacy. Massive stone platforms known as yácatas, curved temple structures unique to Purépecha architecture, overlooked the lake from a high terrace. These yácatas combined rectangular and circular forms, symbolizing cosmic concepts and serving as ceremonial foundations for elite rituals.

Terraced fields, residential neighborhoods, and administrative compounds extended outward from the ceremonial core. Tzintzuntzan became the seat of the cazonci, the Purépecha ruler who presided over an empire that stretched from the Pacific coast to the fertile valleys near present-day Guanajuato.

## Masters of Copper and Iron

One of the Purépecha's most remarkable achievements was their advancement in metallurgy. By the Late Postclassic period, they were the foremost metalworkers in Mesoamerica. Purépecha artisans cast and hammered copper and alloys into axes, bells, needles, fishhooks, jewelry, and ceremonial objects. They also worked with bronze, an alloy that remained rare elsewhere in Mesoamerica.

Metal production centers, particularly in the Zacapu and Pátzcuaro regions, were tied directly to the empire's political authority. The state controlled access to ore sources and coordinated the distribution of metal goods, using them as both tools and symbols of elite power.

This technological strength played a notable role in shaping Purépecha identity and diplomacy. Their metal weaponry and tools gave them tactical advantages over rival states and helped sustain their independence.

## The Empire That the Aztecs Could Not Conquer

In the fifteenth century, as the Aztec Triple Alliance expanded across central Mexico, most neighboring polities were absorbed or subdued. The Purépecha were the exception. Their territory formed a powerful buffer to the west, and repeated Aztec attempts to invade failed.

The Aztec ruler Axayacatl led a major campaign into Michoacán around 1479 and expected victory over what he believed was a lesser power. Instead, his forces encountered well-organized Purépecha armies equipped with metal-tipped weapons and supported by strong defensive terrain. The Aztecs suffered heavy losses and retreated. They never attempted a large-scale invasion again.

This enduring independence allowed the Purépecha to maintain cultural and political autonomy, even as the Aztec Empire reached its height. Trade between the two states did occur, with obsidian, cotton, feathers, and foodstuffs moving across borders, but always under the shadow of mutual wariness.

## Lake Pátzcuaro's Island Cities

Several islands within Lake Pátzcuaro held significant cultural and political importance. At Janitzio, steep terraces climbing the island's slopes once supported temples and elite residences. Other islands, such as Pacanda and Yunuén, played roles in regional trade, fishing economies, and religious ritual.

Lake transport connected these island centers to lakeshore cities and formed an internal network that supported mobility, communication, and defensive strategy. Canoes made from hollowed logs navigated the water and carried fish, ceramics, woven goods, and ceremonial offerings.

The lake's shifting shorelines over centuries have exposed and concealed archaeological remains. Today, stone foundations visible at low water hint at the once bustling settlements that lined the shores.

## Social and Political Organization

The Purépecha governed through a centralized structure centered on the cazonci and a council of nobles. Administrative regions were divided into districts overseen by local leaders who collected tribute, organized labor, and maintained religious institutions.

Religion played a major role in political life. Deities associated with fire, war, and the sun occupied central positions in Purépecha cosmology. Ceremonial fires burned at major temples, and priests performed rituals that reinforced state power and community cohesion.

Unlike some Mesoamerican cultures, the Purépecha relied more heavily on stone temples and public spaces than on carved monuments. Their architecture emphasized platforms, terraces, and circular forms that reflected a worldview rooted in cycles and movement.

## Spanish Contact and the Transformation of Purépecha Cities

When the Spanish arrived in the early sixteenth century, the Purépecha Empire remained strong but not without internal tension. The last cazonci, Tangaxuan II, initially negotiated with the invaders in an effort to protect his people from the violent conquests occurring in the Aztec heartland. However, political conflict and Spanish ambition quickly destabilized the region.

In 1530, conquistador Nuño de Guzmán executed Tangaxuan II and triggered upheaval across the empire. Forced labor, disease, and colonial restructuring disrupted long-established political and economic systems. Many Purépecha cities shrank or transformed under Spanish rule.

Tzintzuntzan remained important as a colonial center but lost its status as an imperial capital. New towns emerged under Spanish administration, and traditional ceremonial practices were suppressed or adapted to survive under new conditions.

Despite these challenges, Purépecha communities preserved elements of their cultural identity, language, and craft traditions, including metalworking, throughout the colonial era and beyond.

## Why the Cities Declined

The decline of Purépecha cities after Spanish contact stemmed from multiple factors.

Disease devastated populations. Like other Indigenous societies, the Purépecha suffered catastrophic losses from smallpox, measles, and other illnesses introduced from Europe.

Political disruption followed the execution of Tangaxuan II. Without centralized leadership, traditional governance structures fractured, which made communities vulnerable to forced resettlement.

Economic shifts redirected labor toward colonial enterprises. Traditional trade networks collapsed, and communal agricultural systems were replaced with Spanish landholding practices.

Religious suppression weakened ceremonial centers. Temples were dismantled or repurposed, and new churches rose atop ancient plazas and platforms.

As a result, many Purépecha urban centers changed dramatically or were abandoned. Yet the cultural landscape of Lake Pátzcuaro remained active, with Indigenous communities adapting to the new conditions while maintaining continuity with ancestral traditions.

## Archaeology and Rediscovery

Systematic archaeological work in Michoacán began in the twentieth century. Excavations at Tzintzuntzan revealed yácatas, ceremonial plazas, and evidence of high-status crafts. Researchers mapped settlement patterns around Lake Pátzcuaro and identified domestic structures, workshops, and shoreline features that illuminated Purépecha daily life.

Studies of metalworking centers uncovered furnaces, slag, molds, and tools that demonstrated advanced metallurgical knowledge. These findings helped clarify why the Aztecs viewed the Purépecha as technologically formidable.

Underwater archaeology in the lake has located submerged structures and provided clues about changing water levels and the early layout of shoreline communities.

As sunlight spreads across Lake Pátzcuaro and illuminates islands that once anchored an empire, the legacy of the Purépecha becomes clearer. Their cities, which included fortified capitals, lakeshore settlements, and island strongholds, defied the power of the Aztec world and flourished through diplomacy, innovation, and metalworking mastery.

Their decline came not through defeat in battle but through the upheaval of colonial transformation, disease, and political manipulation. Yet Purépecha identity endures. Languages, crafts, and community traditions continue in the towns surrounding the lake and echo the strength of a people who built cities in the highlands and resisted one of the greatest empires of Mesoamerica.

The ruins along the shores remain quiet, but they speak of resilience, ingenuity, and an empire that once held the power to stand alone in a world of rising forces.

# Chapter 14
# Lines of Stone — The Geometric Cities of the Ohio Valley

A broad aerial view opens over the Ohio Valley. Green fields stretch into the distance and are divided not by fences or roads but by sweeping arcs and straight lines pressed into the earth. Perfect circles, vast squares, and octagonal forms emerge as light shifts across the landscape. Some are so large they can only be fully grasped from the sky. These geometric earthworks, raised embankments of soil meticulously shaped more than two thousand years ago, were once the ceremonial centers of the Adena and Hopewell cultures. Their scale, symmetry, and engineering precision reveal a world where gathering places took the role of cities and drew people from across eastern North America.

## Landscapes of Ceremony and Power

Between roughly 1000 BCE and 400 CE, the Adena and Hopewell cultures of the Ohio Valley built some of the most complex earthworks in prehistoric North America. Unlike later urban civilizations with dense populations and permanent cities, these societies organized themselves around large ceremonial centers used for gatherings, rituals, trade, and diplomacy.

The Adena, the earlier of the two traditions, developed burial mounds, circular enclosures, and ceremonial landscapes that shaped later Hopewell practices. By the time of the Hopewell tradition, earthworks reached geometric perfection. Embankments formed sprawling squares measuring hundreds of feet on each side, circles more than a thousand feet across, and octagons aligned with astronomical events.

The most iconic complexes, located near present-day Newark, Chillicothe, and Portsmouth, demonstrate remarkable planning. Constructed with simple tools, these earthen monuments required coordinated labor and a deep understanding of geometry, surveying, and astronomy.

## The Geometry of Gathering, Centers Without Walls

Hopewell ceremonial centers were not cities in the conventional sense. Instead, they functioned as vast ritual landscapes visited seasonally by dispersed communities. People lived in small villages scattered across river valleys, yet they traveled to these monumental sites to participate in ceremonies, exchange goods, and strengthen alliances.

The Newark Earthworks illustrate this design on an extraordinary scale. The site included a massive octagon connected to a circular enclosure, a Great Circle with monumental gateways, and a network of parallel embankments. Alignments within the octagon correspond to the moon's 18.6-year cycle and showcase astronomical knowledge far more advanced than once assumed.

At Chillicothe, multiple earthwork complexes, including squares, circles, parallel walls, and burial mounds, formed a ceremonial corridor along the Scioto River. These gatherings created regional identity and shaped cultural life much like cities did elsewhere, even without dense permanent populations.

## Engineering With Earth and Vision

The earthworks required planning techniques that rivaled those of ancient urban societies. Builders excavated soil using wooden spades, stone tools, baskets, and fire-hardened implements. Embankments often rise only a few feet, yet their precision, straight lines, perfect circles, and consistent dimensions, indicates advanced surveying methods.

Recent analyses suggest that Hopewell engineers used cords, sight poles, and horizon markers to lay out geometric plans. The accuracy is striking, with many circles deviating by only a few inches across their entire circumference. The astronomical alignment of the Octagon Earthworks suggests that ceremonial events may have been tied to lunar cycles, agricultural rhythms, or beliefs about cosmic order.

These earthworks were intentional constructs, planned landscapes designed to shape movement, visibility, and the ceremonial experience.

## Trade Networks That Spanned a Continent

Hopewell and Adena ceremonial centers served as exchange hubs for materials gathered from distant regions. Archaeological excavations reveal copper from the Great Lakes, obsidian from the Rocky Mountains, mica from the Appalachian Mountains, marine shells from the Gulf of Mexico and the Atlantic Coast, and grizzly bear teeth from the northern plains. These materials were transformed into intricately crafted objects such as copper axes and breastplates, mica cutouts, obsidian blades, pottery, and carved stone pipes. Many items held ritual significance and were deposited in burial mounds or ceremonial caches.

The enormous reach of this trade network suggests political alliances, communication systems, and shared ceremonial practices that extended across thousands of miles.

## Ritual, Leadership, and Social Life

Leadership within Hopewell societies appears to have been based on ritual authority rather than centralized political control. Burial mounds contain elaborate grave goods that indicate certain individuals held special roles, likely as religious leaders, skilled artisans, or community representatives.

Ceremonies involving feasting, dance, and the exchange of crafted gifts reinforced social bonds. Burial rituals conducted within or near the earthworks honored ancestors and affirmed community continuity. These gatherings created the social fabric of Hopewell life and linked small villages into a broad ceremonial world.

## Decline and Abandonment

By around 400 CE, construction of major Hopewell earthworks slowed dramatically. Over the following centuries, many ceremonial centers were gradually abandoned. Villages remained occupied, but the large-scale gatherings that defined Hopewell life faded.

Researchers propose several possible causes. Environmental pressures may have disrupted food production and reduced the surplus needed for ceremonial gatherings. Shifts in political or religious authority may have weakened the influence of ritual leaders who once coordinated regional events. Communities may have adopted new ceremonial practices that no longer required monumental earthworks. Population dispersal into new river valleys may have shifted settlement patterns enough to make centralized ceremonial hubs less relevant.

By the time European explorers arrived in the region, most earthworks had long since been reclaimed by forests and forgotten by local memory.

## Rediscovery Through Survey and Science

Interest in the Ohio Valley earthworks grew during the eighteenth and nineteenth centuries when European-American settlers encountered massive mounds and embankments that seemed too large to have been built by Indigenous peoples. Early antiquarian writers mistakenly attributed them to a "lost race," ignoring Native oral histories and misinterpreting the archaeological record.

Systematic surveys began in the 1840s with the work of Ephraim Squier and Edwin Davis. Their detailed maps provided some of the earliest scientific documentation of the earthworks and demonstrated their geometric precision.

In recent decades, LiDAR technology has revolutionized understanding of these landscapes. By scanning through vegetation, LiDAR reveals ground features invisible at eye level, including subtle embankments, parallel wall systems, and previously unknown geometric forms. Many sites once thought destroyed or incomplete have reemerged as coherent ceremonial complexes.

Radiocarbon dating, soil analysis, and ground-penetrating radar continue to refine timelines and expose structures buried beneath plowed fields or wooded hillsides. These discoveries confirm that Hopewell and Adena ceremonial centers were not isolated monuments but parts of vast, interconnected landscapes.

From the air, the geometric cities of the Ohio Valley reveal a world built through vision, precision, and ceremony. Circles, squares, and octagons stretch across green fields and stand as reminders of societies that organized life not through dense urban centers, but through monumental gatherings rooted in ritual and community.

The Adena and Hopewell shaped a cultural region defined by astronomy, trade, and shared identity. Their decline remains a quiet mystery, yet their legacy endures in the contours of the land and the insights of modern archaeology.

These lines of stone and earth carry forward the story of a world that understood geometry as sacred, landscape as community, and ceremony as the heart of cultural life. Their cities may have vanished, but their patterns remain etched into the earth, a testament to a people who built with vision, precision, and purpose long before their complexity was fully recognized.

# Chapter 15 The City That Warned the Night Sky — Toltec Ascendancy and the Fall of Tula

Stone warrior columns stand rigid on a hillside, their carved helmets and butterfly breastplates framed against a storm-dark sky. Lightning splits the horizon and illuminates terraces, toppled walls, and the outlines of what was once a commanding ceremonial center. This is Tula, known in ancient stories as Tollan, the capital of the Toltec world. It was a city whose influence shaped Central Mexico long after its decline around 1150 CE, and whose ruins continue to provoke debates about memory, myth, and power.

## The Rise of a Highland Power

Tula developed around 900 CE in the Tula River Valley, a strategic highland basin northeast of modern

Mexico City. After the fall of Teotihuacan centuries earlier, no single state dominated central Mexico. Instead, regional polities competed for authority. Tula quickly distinguished itself through its access to fertile fields, its proximity to obsidian deposits, and its position along emerging trade routes linking the central highlands to northern and western Mexico.

The Toltecs who built and governed Tula did not form an empire in the later Aztec sense, but they exercised regional power through alliances, military presence, and the spread of a distinctive visual language. At its height, the city held tens of thousands of residents. Houses and workshops spread across surrounding hillsides while the ceremonial core, marked by pyramids, plazas, and columned halls, projected unmistakable authority.

## Warrior Imagery and the Architecture of Power

Toltec ideology is most visible in the art and architecture of the ceremonial precinct. The famous Atlantean figures atop Pyramid B, carved as towering warriors with atlatls, spearpoints, and elaborate regalia, illustrate a leadership culture that emphasized martial strength. Relief carvings on walls and staircases depict jaguars, coyotes, and eagles consuming human hearts. These images reflect the fusion of warfare, sacrifice, and cosmic order that defined Toltec identity.

The ceremonial structures themselves reinforced this worldview. Temple-pyramids served as stages for state rituals and elite ceremonies. Colonnaded halls with carved columns housed formal gatherings and symbolically elevated the status of the warrior class. Ballcourts tied athletic contests to diplomacy and ritualized conflict, binding political relationships to sacred performance. Together, these elements created an urban environment where power was communicated through architecture, imagery, and performance.

## Craft Production and Economic Networks

Economic strength supported Tula's political and ceremonial life. The capital hosted specialized workshops that produced obsidian blades, decorated ceramics, woven textiles, shell ornaments, and cast metal objects. Obsidian mining areas nearby supplied raw materials for tools and weapons traded throughout the region. Turquoise, prized for its symbolic value, linked Tula to communities in the American Southwest.

Plumbate ceramics, known for their glossy surfaces and distinctive forms, appeared at Tula and at sites far beyond the Toltec homeland. These ceramics, along with metalwork and carved stone objects, traveled through long-distance networks that tied Tula to central Mexico, northern Mexico, and perhaps even parts of Central America. These connections strengthened the city's regional importance and allowed Toltec symbols to spread widely.

## Influence Across Central Mexico

The influence of the Toltecs remains one of the most debated subjects in Mesoamerican archaeology. Aztec histories written centuries later describe the Toltecs as master builders and wise rulers. The legendary figure Ce Acatl Topiltzin Quetzalcoatl, portrayed as a moral leader forced into exile, became central to these narratives.

Archaeological evidence confirms Tula's prominence but does not support the far-reaching empire described in later accounts. Even so, Toltec architectural forms, warrior imagery, and ceremonial symbols appear at distant sites. Chichén Itzá on the Yucatán Peninsula displays clear affinities with Tula in its colonnaded halls, feathered serpent motifs, and military iconography. Whether these similarities reflect direct contact, migration, shared ideology, or imitation remains debated. What is clear is that the Toltecs shaped a visual and symbolic vocabulary that became widely recognized across Mesoamerica.

## Urban Life in Tula

Tula was a vibrant city as well as a ceremonial center. Archaeology reveals dense residential districts composed of multi-room houses built from stone and adobe. These neighborhoods were arranged around patios used for cooking, storage, and social gatherings. Workshops produced obsidian tools, pottery, textiles, and shell ornaments at a significant scale. Farmers cultivated maize, beans, squash, and maguey using terracing and irrigation systems that helped stabilize food supplies for the growing population.

Daily life in Tula unfolded within neighborhoods tied to specific crafts or social groups. Marketplaces moved goods between urban and rural communities. Roads and pathways linked Tula to nearby settlements that depended on the capital for trade, ceremonial events, and political oversight. These networks allowed the city to function as a regional hub of culture and governance.

## Debates on Toltec Legacy and Identity

The contrast between Aztec narratives and archaeological findings has led to ongoing debates about the nature of the Toltec state. Many scholars now view Aztec descriptions as partly mythologized, created to legitimize Aztec authority by connecting it to an idealized past. Archaeological excavations show a more grounded picture of Tula as a strong regional center with impressive art, architecture, and economic influence, but not the vast empire sometimes described in colonial-era texts.

Yet mythology is itself a powerful form of legacy. The persistence of Toltec imagery in later Mesoamerican art, architecture, and ideology demonstrates the enduring impact of Tula long after its decline. The memory of Tollan became a symbol of ideal rulership, artistic excellence, and political legitimacy.

## Decline Around 1150 CE

Tula's decline around 1150 CE appears to have resulted from multiple pressures that unfolded gradually. Paleoclimate evidence from lake sediments and tree-ring records indicates severe drought across central Mexico during the eleventh and twelfth centuries. These prolonged dry periods strained agricultural production and reduced the surplus needed to sustain urban life.

Archaeological traces suggest internal unrest. Several ceremonial buildings show fire damage, abrupt abandonment, or incomplete construction phases. These signs point to factional conflict or political instability that weakened central authority. Shifting trade routes may also have undermined the economic foundation of the city, especially if access to obsidian or turquoise diminished.

External pressures, including migration or conflict with northern groups, may have contributed, though evidence is limited. What is clear is that by the late twelfth century, residential districts were largely abandoned and the ceremonial center no longer served as a political capital. When the Aztecs rose centuries later, they encountered Tula as a ruin filled with symbolic power.

## Archaeology and the Reconstruction of Tula

Excavations throughout the twentieth and twenty-first centuries uncovered Tula's monumental core, including Pyramid B, the Atlantean figures, the ballcourt, and the colonnaded halls. Researchers identified residential districts that reveal the rhythms of daily life, including crafting areas, food production sites, and domestic spaces. Water management systems, including channels and reservoirs, illustrate the engineering that sustained the city.

Modern mapping technologies such as LiDAR have revealed buried foundations and topographic details that expand the known footprint of ancient Tula. Comparative studies with Teotihuacan, Chichén Itzá, and later Aztec sites have deepened understanding of architectural continuity and symbolic transformation. These discoveries help clarify Tula's role as a bridge between earlier urban traditions and the imperial ideologies that shaped later Mesoamerican history.

Lightning fades across the hillside as the warrior columns return to stillness, their stone faces turned toward the valley. Tula's rise and fall reveal the delicate balance between power, environment, and human ambition. The city projected strength through its architecture, ritual imagery, and craft production, shaping regional culture and leaving a symbolic legacy that endured long after its decline.

Tula's fall around 1150 CE reminds modern readers that even the most influential cities rely on fragile systems, including climate stability, political unity, and economic networks, that can shift unexpectedly. Yet its lasting presence in memory and myth illustrates the enduring resonance of human creativity. The city that once warned the night sky with its warrior columns now stands as a testament to the epics, achievements, and vulnerabilities that define all civilizations.

# Chapter 16
# A Kingdom of Wood and Rain — The Amazon's Hidden Urban Grid

Sunlight breaks through the dense Amazon canopy in narrow, golden beams. Leaves shimmer, insects hum, and the forest floor remains concealed beneath a thick tapestry of green. Yet from above, where satellites and LiDAR sweep across northern Bolivia and western Brazil, geometric patterns suddenly emerge. Circles, squares, causeways, and earthen mounds form a gridlike network across the jungle. Beneath the world's largest rainforest lie the traces of settlements, ceremonial centers, and agricultural landscapes that challenge long-held assumptions about Amazonian history. This was not an untouched wilderness, but a kingdom of wood and rain, an engineered environment shaped by societies whose achievements remained hidden for centuries.

## Revealing a Lost World Beneath the Trees

For generations, scholars believed the Amazon lacked the conditions necessary for large urban centers. Nutrient-poor soils, dense vegetation, and seasonal flooding seemed to limit sustained population growth. Indigenous communities were thought to live primarily in small, scattered groups. Yet beginning in the late twentieth century, a wave of discoveries overturned this belief.

The first hints came from the geoglyphs of Acre, enormous earthwork enclosures revealed when forest was cleared for ranching. Geometric forms hundreds of feet across suggested a sophisticated ceremonial landscape long predating European arrival. LiDAR surveys expanded the picture even further. Hidden beneath dense canopy, researchers found interconnected settlements with causeways, defensive ditches, plazas, and raised agricultural fields.

Instead of isolated villages, many Amazonian societies lived in low-density urban networks, with clusters of engineered settlements forming large regional systems. These communities shaped the landscape to control water, improve soils, and support growing populations, revealing a level of complexity once deemed impossible.

## Fortified Settlements and Regional Systems

One of the most significant areas of discovery lies in the southwestern Amazon, where a network of Late

Pre-Columbian earthwork cities associated with the Casarabe culture flourished between 500 and 1400 CE. Though made primarily of earth and wood, their settlements show urban planning comparable to other early civilizations.

Large centers such as Cotoca and Landívar contain platform mounds, ceremonial plazas, rectangular and circular enclosures, and gridded causeways linking them to surrounding communities. Deep ditches often encircle these sites, marking defensive concerns or sacred boundaries. Elevated roads stretch outward for miles, uniting smaller settlements into a cohesive region.

Excavations reveal permanent populations engaged in farming, fishing, craft production, and trade. Wooden structures once shaped these plazas and platforms, though little remains due to tropical decay. Stone was scarce, but earth served as a versatile medium for architecture, sculpted into mounds, canals, and embankments.

## Raised-Field Agriculture: Engineering the Flooded Plains

Agriculture in the Amazon operated on principles distinct from irrigation or highland terrace systems. Many societies relied on raised-field agriculture, constructing long, elevated ridges above seasonal flood zones. Channels between fields controlled water levels, prevented crop damage, and created habitats for fish and aquatic plants.

In the Llanos de Moxos of Bolivia, archaeologists have mapped tens of thousands of raised fields across the savanna wetlands. These fields are accompanied by causeways, canals, fish weirs, and artificial ponds. Together, they formed a productive agricultural system capable of supporting substantial populations.

The labor required to construct and maintain these earthworks suggests organized community effort and long-term planning. The extent of the system shows that Amazonian societies did not merely adapt to their environment. They actively reengineered it.

## Urbanism Without Stone

Unlike the stone cities of the Andes or Mesoamerica, Amazonian urban centers were built largely from wood, earth, and cane—materials that decay rapidly in tropical climates. As a result, their architectural legacy appears subtle, preserved mainly through earthworks, soil chemistry, and traces of postholes.

Ceremonial mounds and plazas served as symbolic and physical centers of community life. Elevated walkways connected neighborhoods, while engineered ponds and reservoirs managed water throughout the year. The Xingu region of central Brazil reveals clusters of circular villages arranged around plazas and connected by straight roads lined with palisades. These linked settlements formed regional systems that may have supported populations similar to many ancient cities worldwide.

## Challenging Old Assumptions

Discoveries from the past two decades show that the Amazon was not a pristine wilderness before European contact, but a landscape shaped by human engineering. Many communities created terra preta, a nutrient-rich artificial soil composed of charcoal, organic waste, and pottery fragments. This dark earth supported stable agriculture in areas with naturally poor fertility.

Road networks reveal large-scale planning; settlements display spatial hierarchy; and the scale of raised fields, canals, and earthworks demonstrates societies capable of mobilizing labor across regions. These findings challenge earlier assumptions that the Amazon lacked the capacity for complex urban development and reveal a mosaic of interconnected systems adapted to rainforest ecology.

## Decline, Depopulation, and the Impact of Climate

The decline of Amazonian urban systems unfolded gradually and varied across regions. Climate records indicate that parts of the southwestern Amazon experienced severe drought episodes between 900 and 1200 CE. These fluctuations disrupted water management systems, reduced crop yields, and weakened settlement stability. Increased fire activity may have further damaged the landscape.

Some regions show signs of internal reorganization, with settlements shifting locations or decreasing in scale as environmental and social pressures mounted. Yet many systems endured for centuries until a far more devastating force reshaped the continent: European contact.

The introduction of Old World diseases, including smallpox, measles, and influenza, caused catastrophic population loss. Mortality rates often exceeded 80 percent. Entire communities vanished within generations, leaving earthworks abandoned and agricultural fields reclaimed by forest. This sudden collapse erased much of the cultural knowledge required to maintain engineered landscapes.

By the time European explorers ventured into the interior, the vast earthwork networks had disappeared beneath dense vegetation.

## Archaeology and Rediscovery

The rediscovery of Amazonian urbanism unfolded slowly. Nineteenth-century travelers noted scattered mounds and embankments, but the full extent remained unknown. Aerial photography in the twentieth century offered glimpses, while forest clearing revealed geoglyphs that renewed scientific interest.

LiDAR transformed understanding in the twenty-first century. By penetrating the forest canopy, it revealed expansive settlement networks, transportation grids, and engineered landscapes. Excavations confirmed these findings through ceramics, tools, hearths, and modified soils that testified to long-term habitation.

What once seemed isolated anomalies now form a vast archaeological record of complex societies hidden beneath rainforest cover.

As sunlight cuts through the canopy and geometric patterns appear beneath the trees, a new view of the Amazon comes into focus. These earthworks and settlement networks reveal a civilization built not on stone temples or imperial conquest, but on wood, water, earth, and collective ingenuity. Their cities blended into the forest; their agriculture responded to seasonal cycles; their roads connected communities in ways now only faintly visible.

The catastrophic depopulation following European contact allowed the forest to reclaim engineered landscapes with silent efficiency. Yet their rediscovery reshapes our understanding of Amazonian history, proving that the region once supported complex urban societies adapted to one of the world's most challenging environments. This hidden kingdom of wood and rain stands as a testament to human resilience, environmental mastery, and the enduring ability of archaeology to rewrite the past.

# Chapter 17
# Fire in the River Valleys — The Collapse of the Mississippian South

Smoke drifts across a broad southeastern horizon, rising from controlled burns set to renew modern farmland. As it settles, the silhouettes of ancient mound centers emerge, their flat-topped pyramids rising from river valleys and their edges softened by centuries of erosion. Once, these mounds anchored powerful cities filled with plazas, wooden temples, palisaded defenses, and trade networks stretching from the Gulf Coast to the Great Plains. Today they sit in quiet fields, reminders of the Mississippian civilizations that flourished for centuries before unraveling in a storm of warfare, disease, and upheaval that reshaped the American South.

## Cities Beyond Cahokia

While Cahokia often dominates Mississippian history, many other urban centers rose across the southeastern United States after 1000 CE. These cities developed their own political systems, ceremonial complexes, and artistic traditions, yet all shared the Mississippian foundations of platform mounds, ritual plazas, and hierarchical leadership.

Etowah in present-day Georgia became a major political and ceremonial center. Its great mounds overshadowed plazas, residential districts, and defensive ditches. Copper plates and shell carvings found in its burials reveal a society tied to the Southeastern Ceremonial Complex, a shared iconographic tradition emphasizing warriors, mythic beings, and cosmic imagery.

Farther west, Moundville in Alabama grew into a city of more than twenty platform mounds arranged around a central plaza. Archaeology suggests a tiered society led by ritual specialists and supported by artisans who produced pottery, shell ornaments, and engraved stone objects.

To the north and west, Spiro Mounds in eastern Oklahoma became one of the most extraordinary ceremonial centers in North America. Its burial deposits contained engraved shell cups, copper plates, carved effigies, beads, and textiles linking the region to distant cultures. Spiro functioned as a sacred gathering place for centuries, drawing populations from across the Mississippi Valley and southern plains.

Together, these cities formed a constellation of Mississippian power, each governing its own territory while interacting through diplomacy, trade, and spiritual exchange.

## Political Structures and Ritual Authority

Mississippian societies were organized around chiefdoms ruled by hereditary elites who held both spiritual and administrative authority. Chiefs often claimed descent from ancestral or mythic beings. They directed ceremonies, mediated alliances, and organized labor for mound construction, agricultural coordination, and public works.

Platform mounds symbolized this leadership. Each supported buildings where elites lived, governed, or conducted rituals. Plazas served as public stages for ceremonies that reinforced social identity and strengthened political hierarchy. Craft specialists produced copper plates, shell carvings, stone effigies, and ceremonial pottery used in diplomacy and ritual exchange.

Warfare played a central role in maintaining authority. Fortified towns, palisaded walls, and burned villages indicate conflict among competing chiefdoms. Military success enhanced a leader's prestige, while failure destabilized entire regions.

## A Continental Web of Trade

Mississippian trade networks stretched across much of North America. Shell from the Gulf Coast, copper from the Great Lakes, mica from the Appalachians, and chert from the central plains moved through exchange routes connecting local farmers to distant ceremonial centers.

At Spiro, engraved shell cups depict cosmological scenes of warriors, mythic creatures, and layered worlds. Copper plates from Etowah portray figures associated with the Southeastern Ceremonial Complex. These objects demonstrate a shared symbolic system that linked communities over vast distances.

Trade brought social prestige and spiritual significance. Yet reliance on imported materials made chiefdoms vulnerable to disruptions in political relationships and shifting regional power.

## Stress Before Contact

By the fifteenth century, signs of strain appeared in several Mississippian regions. Environmental changes tied to the Little Ice Age affected agricultural yields. Population growth increased pressure on farmland. Competition for resources intensified conflict.

Burned villages, fortified towns, and shifting settlement patterns reveal a landscape marked by instability. Some mound centers, like Etowah, experienced cycles of abandonment and reoccupation. Others, such as Moundville, lost influence as populations dispersed to surrounding areas.

Despite these pressures, Mississippian communities remained resilient. They continued to farm, trade, and conduct ceremonies. Their ultimate collapse would come not from internal challenges alone, but from forces they could not anticipate.

## A New and Devastating Kind of Fire

European arrival in the sixteenth century unleashed catastrophic disruptions across the Mississippian world. Diseases such as smallpox, measles, influenza, and bubonic plague spread along trade routes faster than European explorers themselves could travel. Mortality rates soared. Some communities may have lost more than half their population in a matter of years.

These "virgin-soil epidemics" devastated political and religious systems. Chiefs and ritual leaders died suddenly. Ceremonial calendars collapsed. Agricultural cycles faltered as villages struggled to maintain labor forces.

Social upheaval followed. Survivors reorganized into smaller communities or abandoned long-standing settlements. Warfare intensified as weakened chiefdoms competed for resources and stability.

The expedition of Hernando de Soto in the 1540s introduced further chaos. His army moved through the Southeast demanding tribute, taking hostages, and provoking violent conflict with numerous Mississippian towns. Although the expedition failed to establish permanent control, the destruction it caused accelerated the fragmentation of regional alliances and leadership structures.

## Transformation of a Cultural Landscape

In the generations following de Soto's march, the Mississippian South underwent profound transformation. Once-powerful cities faded into memory. Chiefdoms dissolved as survivors formed new alliances within a landscape reshaped by depopulation, political vacuum, and the spread of European goods and rivalries.

Descendant nations such as the Choctaw, Chickasaw, Creek, and Caddo developed new political structures adapted to colonial pressures. They preserved agricultural traditions, ceremonial practices, and aspects of Mississippian cosmology, yet their societies no longer resembled the mound-centered chiefdoms that had existed only a century earlier.

By the eighteenth century, monumental centers like Etowah, Moundville, and Spiro had long been abandoned. Forests grew over their mounds, obscuring the scale of the civilizations that once flourished there.

## Rediscovery Through Archaeology

Archaeological research in the twentieth century renewed attention to Mississippian mound centers. Excavations uncovered ceremonial objects of copper, shell, stone, and pottery that revealed vibrant artistic and ritual traditions. Burial structures and domestic areas offered new insights into social hierarchy, daily life, and spiritual belief.

Today, archaeologists combine excavation with remote sensing, soil analysis, and regional surveys to reconstruct Mississippian networks. Their findings show that Mississippian cities played vital roles in shaping the cultural landscape of North America.

The scale and sophistication of Mississippian civilization were far greater than early researchers realized. Mound complexes now documented across the Southeast reveal centuries of ingenuity, adaptation, and cultural richness.

As smoke drifts across the modern horizon and the ancient mound centers of the Southeast stand in stillness, the broader story of the Mississippian South becomes clear. These were cities of earth and timber shaped by ceremony, trade, leadership, and a vast web of alliances. Their collapse came not from a single cause, but from cascading pressures including environmental stress, internal conflict, devastating disease, and the upheaval brought by European intrusion.

What remains today are mounds, plazas, and artifacts quietly embedded in the landscape. They whisper of a world once vibrant with life and remind modern readers of both the resilience and fragility of human societies layered beneath the river valleys and forests of the American South.

# Chapter 18
# Citadels in the Clouds – The Chachapoya Cities of Northern Peru

Clouds drift low across a narrow Andean ridge, wrapping the slopes in shifting veils of mist. As the fog thins, massive stone walls emerge, tall curved fortifications rising from the cloud forest with their silhouettes suspended above deep valleys carved by ancient rivers. Ferns and orchids cling to moss-covered stones. The air is cool and heavy with moisture, and every sound echoes across steep cliffs. These are the remnants of the Chachapoya, a people who built citadels in the clouds and shaped one of the most dramatic cultural landscapes of ancient Peru.

## A Civilization on the Edge of the Andes

The Chachapoya, often called the "Warriors of the Clouds," inhabited the eastern Andean slopes of northern Peru between roughly 800 and 1500 CE.

Their homeland sat at the meeting point of high mountains and Amazonian cloud forest, a rugged region of sheer cliffs, narrow ridges, and deep valleys that offered both isolation and strategic advantage. Here, they constructed fortified settlements, agricultural terraces, ceremonial plazas, and necropolises that clung to improbable heights.

Unlike coastal civilizations that relied on desert irrigation systems or highland empires that governed broad plains, the Chachapoya lived in a fragmented landscape. Their settlements formed a mosaic of independent or loosely allied communities, each controlling its own ridge, valley, or plateau. Despite these challenges, they developed distinctive architectural and artistic traditions that set them apart from other Andean groups.

## Kuélap: A Fortress Above the Clouds

The most famous Chachapoya site is Kuélap, a massive hilltop fortress perched more than nine thousand feet above sea level. Its outer walls, some rising more than sixty feet high, encircle an urban core filled with hundreds of circular stone structures, ceremonial spaces, and administrative compounds. The narrow, funnel-shaped entrances may have served defensive functions, forcing intruders into single-file lines.

Kuélap's position provides sweeping views of surrounding valleys and early warnings of approaching enemies. Yet the site was not solely a military stronghold. Excavations reveal domestic residences, storerooms, and ritual buildings decorated with geometric friezes. These patterns, including diamonds, zigzags, and anthropomorphic symbols, reflect Chachapoya artistic identity and appear across settlements in the region.

## Hilltop Fortresses and Defensive Networks

Kuélap was only one among many fortified Chachapoya settlements. Hilltop citadels such as Revash, Purunllacta de Soloco, Pajáten, and La Jalca formed a defensive system stretching across the cloud forest. Many fortresses occupy ridges so narrow that a single wall could block access. Others incorporate sheer cliffs into their defenses, creating formidable strongholds.

Circular houses, typically made from limestone blocks, served as the basic residential form. Some included raised platforms, decorative friezes, or interior niches. Agricultural terraces carved into mountain slopes produced maize, potatoes, and quinoa. Forested zones supported hunting and gathering, while carefully managed springs and rainfall provided essential water sources on exposed ridges.

## Burial Sites on Cliff Faces

The Chachapoya developed striking and distinctive burial traditions. Cliff-side mausoleums and anthropomorphic sarcophagi overlook deep gorges, their painted faces gazing outward from ledges high above the valley floor. These sites, including the renowned Revash and Karajía sarcophagi, contain mummies placed within elaborate structures.

The sarcophagi stand several feet tall, shaped like human figures with expressive faces and geometric motifs painted in red, yellow, and black. Their placement on inaccessible cliffs honored ancestors while protecting remains from looting or disturbance.

Nearby cave cemeteries contain mummy bundles wrapped in textiles and accompanied by ceramics or ritual offerings. These practices highlight a deep connection between ancestors, the landscape, and community identity.

## Conflict with the Expanding Inca Empire

In the late fifteenth century, the Chachapoya homeland came under threat from the expanding Inca Empire. The Inca considered the cloud-forest region strategically vital for securing eastern trade routes and preventing rival powers from gaining footholds on the highland borderlands.

The Chachapoya resisted fiercely. Several campaigns under the Inca rulers Tupac Inca Yupanqui and Huayna Capac met with prolonged opposition from fortified cloud-forest settlements. Despite their rugged terrain and strong defenses, the Chachapoya were eventually conquered. Many communities were forcibly resettled under the Inca policy of mitmaqkuna, which aimed to break resistance and integrate diverse groups into the empire.

Even under imperial rule, the Chachapoya retained aspects of their identity. Their skills as warriors and builders were valued, and some groups were relocated to serve military or labor roles in distant regions. Yet traditional governance and social structures were disrupted.

## The Spanish Arrive in a Fractured World

When the Spanish entered the Andes in the 1530s, the Inca world was already destabilized by civil war and epidemic disease. The Chachapoya found themselves caught between a collapsing empire and an unfamiliar colonial force.

Many Chachapoya communities initially supported the Spanish as potential liberators from Inca domination. This alliance, however, quickly deteriorated. European diseases, including smallpox and influenza, swept through the cloud-forest valleys, causing catastrophic population loss. Entire communities were devastated within a few generations.

Spanish demands for tribute, forced labor, and religious conversion further destabilized traditional life. Colonial towns replaced ancestral settlements, and severe population decline made it impossible to maintain terraced fields, fortresses, and ceremonial sites. By the seventeenth century, many Chachapoya centers were abandoned or absorbed into colonial administrative structures.

## Why So Many Settlements Were Abandoned

The abandonment of Chachapoya settlements resulted from intertwined pressures. Population decline due to epidemic disease sharply reduced the labor needed to sustain agricultural terraces and defensive structures. Inca resettlement policies had already weakened local authority, and Spanish colonial demands fragmented communities even further.

Environmental conditions contributed as well. Cloud-forest vegetation quickly overtakes unmaintained constructions. Moss, rain, and shifting soils collapse walls and erase plazas within decades. Over time, even monumental fortresses such as Kuélap became hidden beneath dense foliage.

## Rediscovery in the Modern Era

Chachapoya cities remained largely unknown to the outside world until the nineteenth century, when explorers documented ridge-top ruins and cliff tombs. Local communities, however, had long been aware of these sites, sometimes incorporating them into regional stories.

Scientific archaeological investigation began in the twentieth century and accelerated in recent decades. Excavations at Kuélap uncovered ceremonial structures, domestic buildings, and evidence of long-term occupation. Studies at Revash, Karajía, and other burial sites illuminated funerary customs, artistic traditions, and ancestral veneration.

Airborne surveys, satellite imagery, and remote sensing have revealed dozens of additional settlements hidden beneath cloud-forest canopy, demonstrating that Chachapoya civilization was far more extensive and interconnected than once believed.

As the clouds shift and Kuélap's stone walls rise above the mist, the story of the Chachapoya becomes clearer. They built citadels in the sky, fortified communities that blended architecture, landscape, and spiritual meaning in ways unique to the Andean world. Their resistance to the Inca, their struggle under Spanish colonization, and the devastation brought by unfamiliar diseases shaped a turbulent history that left many of their cities silent.

Yet those cities endure. Their ruins, rediscovered among ferns and orchids, reveal the resilience of a people who lived on the edge of the Andes and created a cultural world as dramatic as the landscape that sustained them. The citadels in the clouds remain as reminders of both the heights of human ingenuity and the fragility of societies confronted by forces beyond their control.

# Chapter 19
# Earthquakes, Empire, and Upheaval — The Fall of Quito's Ancient Cultures

Tremors ripple across Ecuador's highland valleys, sending vibrations through steep slopes and stone foundations. Ancient walls, long abandoned, shiver under the force and collapse into dust. The Andes are restless, shaped by volcanoes, tectonic clashes, and shifting soils that have influenced human life for millennia. In this seismic world, long before the rise of the Inca and far before the arrival of the Spanish, the Quitu and later the Cara peoples built settlements, ceremonial centers, and trade networks that flourished across the northern highlands. Their cultures endured cycles of prosperity and disruption, only to vanish through a combination of natural disaster, political upheaval, and imperial expansion that obscured their histories for centuries.

## Highland Worlds Before Empire

The region known today as Quito sits within a valley framed by volcanoes and crossed by ancient travel routes that once connected the tropical lowlands to the high Andean plateaus. For thousands of years, its fertile terraces, mild climate, and strategic position drew communities that forged distinct cultural identities long before imperial unification.

Among the earliest of these were the Quitu, pre-Inca peoples who occupied the valley and surrounding hills between roughly 800 and 1200 CE. Their settlements varied in size, from hamlets perched on ridges to larger communities arranged around open plazas. Archaeology reveals domestic structures built from earth, stone, and wood, along with agricultural terraces that made efficient use of volcanic soils.

The Quitu organized themselves into a network of local chiefdoms tied by shared beliefs, trade relations, and occasional alliances. They maintained autonomy under their own leaders and adapted to microclimates shaped by dramatic changes in elevation across northern Ecuador.

## Trade Networks Across Mountains and Forests

Trade shaped Quitu economic and social life. The region lay at a crossroads between the northern Andes, the Pacific coast, and the Amazon basin. Goods flowed in every direction. From the coast came Spondylus shells and dried fish. From the highlands came textiles, obsidian blades, and agricultural products. From the eastern forests came feathers, medicinal plants, and resins used for ritual and exchange.

Ceramics, stone carvings, and burial offerings reveal a society connected to distant cultures. Quitu traders moved along mountain trails and river valleys, exchanging materials and ideas with communities as far north as Colombia and as far south as central Ecuador. These routes became pathways for artistic styles, ceremonial traditions, and shared technologies.

## Ceremonial Centers and Sacred Landscapes

Though not urban in the dense sense of later Andean capitals, Quitu communities built ceremonial centers that served as focal points for ritual life. These sacred spaces were often placed on elevated terraces or ridges overlooking valleys—locations chosen for their visibility, symbolism, and connection to surrounding mountains.

Mountain worship played a central role in Quitu and broader Andean cosmology. Nearby volcanoes such as Pichincha, Cotopaxi, and Cayambe shaped communal beliefs as sources of both life-giving water and destructive force. Ritual offerings near springs, caves, and highland passes reveal a spiritual relationship with the landscape viewed as both nurturing and perilous.

Some sacred sites later absorbed into Inca and colonial structures preserve Quitu foundations beneath newer layers. The stratigraphy of Quito's historic center reflects a ceremonial tradition that began centuries before imperial rule.

## The Arrival of the Cara: A New Political Horizon

Around the twelfth century, a new cultural and political presence emerged: the Cara. Believed to have originated from coastal or northwestern regions, the Cara entered the Quitu valleys and gradually asserted dominance. Instead of erasing Quitu identity, they appear to have merged with it, forming a hybrid political entity remembered in later chronicles as the Quitu-Cara or Kingdom of Quito.

Under Cara leadership, regional chiefdoms became more integrated. Oral traditions, though filtered through colonial interpretation, speak of rulers whose influence extended across northern Ecuador. Archaeological evidence supports increased settlement organization and political consolidation during this period.

New fortified hilltops, administrative centers, and agricultural expansions suggest heightened complexity. The political landscape shifted from localized chiefdoms to broader federations that interacted with neighboring highlands and lowlands.

## Landscapes in Motion: Earthquakes and Volcanic Upheaval

While Quitu-Cara institutions grew stronger, the land beneath them remained unstable. The northern Andes sit along a major tectonic boundary, and the region has endured centuries of earthquakes, landslides, and volcanic eruptions. These ongoing natural forces repeatedly reshaped the landscape and disrupted human settlement.

Geological and archaeological evidence shows that powerful earthquakes struck the Quito region during the Late Intermediate Period, damaging structures, destroying terraces, and forcing communities to rebuild or relocate. Stone walls collapsed, irrigation channels broke apart, and agricultural systems faltered under shifting soils.

Volcanic activity added further strain. Ashfall from nearby volcanoes blanketed crop fields, contaminated rivers, and obscured sunlight. Layers of volcanic deposits preserved in archaeological sites reveal cycles of abandonment, recovery, and adaptation.

Natural disasters did not alone end Quitu and Cara culture, but they weakened political cohesion and strained resources at moments when regional competition was intensifying.

## The Andean World Before the Inca

By the early fifteenth century, the Quitu-Cara polity governed a significant portion of northern Ecuador. Its capital, likely near modern Quito, oversaw networks of regional lords, agricultural zones, and trade routes. Warriors defended ridge-top fortresses. Ceremonial centers upheld political legitimacy. Trade caravans moved between coast, highlands, and rainforest.

Yet this system soon faced a far more expansive force. The Inca Empire, expanding northward under rulers such as Tupac Inca Yupanqui and Huayna Capac, sought to bring the northern highlands under imperial rule. The region offered fertile valleys, natural defenses, and strategic access to northern trade routes. It also posed a formidable challenge, as Quitu-Cara leadership commanded respect and loyalty across the valleys.

## Conquest, Resistance, and Imperial Transformation

The Inca launched multiple campaigns to subdue the Quitu-Cara territories. Battles unfolded across high passes, ridge-top forts, and ceremonial centers. Some communities submitted quickly; others resisted fiercely. Chroniclers later described long conflicts that tested Inca endurance.

Ultimately, sustained pressure allowed the Inca to prevail. They reorganized the region under provincial administrators, replaced local leaders, and introduced imperial road networks, storehouses, and ceremonial structures. Mitmaqkuna, colonists loyal to the Inca, were relocated into the region to secure imperial control.

Local customs persisted, but Quitu and Cara political systems were absorbed into the administrative structure of Tawantinsuyu. The region became an important northern frontier of the empire.

## Spanish Arrival and the Final Unraveling

When Spanish conquistadors arrived in the 1530s, the northern highlands were destabilized by Inca civil war between Atahualpa and Huáscar. This turmoil, combined with the rapid spread of European diseases, weakened regional resilience.

Smallpox, influenza, and other epidemics devastated highland populations. Leadership collapsed, labor systems broke down, and agricultural terraces were abandoned. Spanish forces capitalized on these conditions, forging alliances with some groups while violently subjugating others.

Colonial rule reorganized settlements into new towns, replaced ceremonial centers with churches, and redirected labor toward mines and colonial estates. These changes fractured the remaining cultural structures of the Quitu and Cara. Over time, their distinct identities blended into the broader colonial world.

## Forgotten and Rediscovered

For centuries, the pre-Inca past of Quito remained obscured. Spanish chronicles elevated Inca history and colonial narratives while minimizing earlier traditions. Yet beneath modern Quito and across surrounding highlands, remnants of Quitu and Cara life persisted.

Archaeological excavations in the twentieth and twenty-first centuries uncovered architectural foundations, terraces, ritual spaces, and ceramics that revealed a deeper history. Sites such as Rumicucho, Tulipe, and Cochasquí illuminated pre-Inca ceremonial life and architectural innovation.

Cochasquí's monumental earthen pyramids demonstrate engineering sophistication and ritual significance. Tulipe's stone pools, arranged with geometric precision, reveal water management systems and astronomical alignments. These findings confirm that the region's early cultures were organized, adaptive, and deeply connected to broader Andean networks.

As tremors shake the highland valleys and ancient walls crumble into dust, the story of the Quitu and Cara becomes clearer. Their world was shaped by mountains that moved, empires that rose and collapsed, and trade routes that carried ideas as readily as goods. They built ceremonial centers on ridges overlooking volcanic landscapes, navigated political rivalries, and endured natural disasters that challenged their resilience again and again.

Their disappearance unfolded not as a sudden collapse but as a long transformation driven by earthquakes, imperial conquest, and the devastating arrival of European disease and colonial rule. Yet their legacy survives in the landscapes of northern Ecuador—in the ruins that rise from Andean terraces, in the cultural memory of descendant communities, and in the foundations of Quito itself.

The ancient cultures of the highland valleys did not vanish without leaving traces. Their stones still quiver when the earth trembles, reminding modern observers that entire civilizations once thrived in the shadow of mountains that could shake their world apart.

# Chapter 20
# Epilogue:
# The Echoes That Remain — How Lost Cities Continue to Shape America's Past

Abandoned plazas dissolve into view, wide and empty where thousands once gathered beneath wooden roofs and rising platforms. Silent pyramids stand against drifting clouds, their edges softened by centuries of wind and rain. Cliffside dwellings cling to canyon walls, their doorways opening onto worlds long vanished. Forest-covered ruins blur into rolling suburbs and the steel outlines of modern skylines. Across this shifting montage, past and present fold into one another, revealing a landscape shaped not only by what survives aboveground, but by the vast, unseen foundations of civilizations that flourished, faltered, and disappeared across the Americas.

## The Ancient Cities Beneath Our Feet

The cities explored throughout this book, from Cahokia and Teotihuacan to Tikal, Mesa Verde, Moundville, and Kuélap, were not footnotes to world history. They were vibrant centers of innovation shaped by environmental challenges, cultural ingenuity, and political ambition. Their plazas once echoed with trade, ceremony, diplomacy, and conflict. Their architecture expressed worldviews that understood mountains, rivers, and stars as active forces in daily life.

Many of these cities thrived for centuries before collapsing or transforming under pressures that remain familiar today. Climate shifts, resource shortages, political upheaval, and external intrusion shaped their destinies. Their ruins inspire modern observers not only to ask why they fell, but how they rose, and how people across the Americas engineered solutions long before European contact.

## Innovation Rooted in Landscape

A powerful theme running through these vanished worlds is the relationship between people and place. The ancient Americas hosted rainforests, deserts, mountains, river basins, and plains. Each environment shaped distinct forms of urban life.

In the Mississippi Valley, societies engineered massive earthworks and ceremonial centers. In the Amazon, communities built raised fields, ditches, and causeways to manage floodplains and support sprawling, low-density urban networks. In the Andes, cultures like the Chachapoya carved fortresses into cloud-forest ridges, transforming extreme terrain into defensible and productive landscapes.

These cities reveal people who responded to their environments with creativity rather than submission. Their innovations show deep ecological knowledge, a quality as relevant to modern challenges as it was thousands of years ago.

## Resilience and Adaptation

Across the Americas, societies confronted droughts, floods, volcanic eruptions, and unpredictable climate cycles. Many responded through technical adaptation or social reorganization. Irrigation systems expanded. Terraces multiplied. Trade networks shifted. Leadership structures changed.

Yet resilience has limits. When environmental stress, political fragmentation, or resource strain intensified beyond what systems could absorb, cities weakened or dispersed. Classic Maya kingdoms, Mississippian chiefdoms, and Andean polities all experienced decline shaped by these pressures.

Still, collapse did not erase legacy. Descendant communities preserved cultural knowledge, adapting ancient lifeways to new conditions. Their survival reveals a continuity that runs through the archaeological record and into living traditions.

## Power, Memory, and Political Change

Urban centers across the Americas were deeply political. They concentrated authority, coordinated labor, and shaped regional identity. When leadership faltered or external forces intruded, cities shifted accordingly.

The Purépecha capitals resisted Aztec expansion. Tula shaped a military and artistic legacy that spread far beyond its boundaries. The Quitu and Cara constructed fortified landscapes that later became battlegrounds for Inca expansion.

Memory also played a transformative role. Abandoned cities often became sacred spaces, woven into myth and legend. Teotihuacan inspired reverence for centuries after its fall. Later peoples understood its ruins as realms of gods and ancestors, demonstrating the power of urban landscapes to shape imagination long after their political institutions dissolved.

## The Fragility of Human Systems

Many ancient American cities rose to great heights only to decline when foundational systems faltered. Agriculture, trade, leadership, and ritual all balanced the needs of densely organized populations. When droughts lasted too long, when warfare intensified, or when pandemics shattered communities, these networks unraveled.

Their ruins remind modern societies that complexity does not guarantee longevity. Urban life depends on delicate systems of cooperation, adaptation, and shared purpose. When those systems fail, cities become vulnerable.

Yet disappearance is rarely total. Many Indigenous communities today maintain traditions that reflect ancient landscapes, rituals, and understandings of place. Their continuity challenges the idea of collapse as an ending, revealing instead the evolution of cultural memory.

## Archaeology's Pursuit of Truth

The rediscovery of these worlds depends on scientific tools and persistent curiosity. Archaeology has progressed from early mapping and excavation to advanced technologies that reveal entire civilizations hidden beneath forests, deserts, and fields.

LiDAR reveals geometric earthworks and settlements long masked by vegetation. Drone imaging uncovers ancient road systems. Soil chemistry, pollen studies, and isotopic analysis reconstruct diets, migrations, and climate patterns. These tools allow researchers to question earlier assumptions and recover histories that colonial narratives overshadowed.

Descendant communities play a vital role, contributing oral memory, cultural insight, and local knowledge that enrich interpretation and safeguard heritage.

## A New Understanding of America's Past

The lost cities of the Americas reshape global history. They demonstrate that complexity, innovation, and urbanism were not exclusive to the Old World. American civilizations developed independently, creating monumental architecture, sophisticated agriculture, and intricate political systems that rivaled any across the globe.

Their achievements expand the understanding of human ingenuity. They reveal cycles of growth, reinvention, and adaptation shaped by spiritual beliefs, environmental knowledge, and social collaboration. They challenge the idea of progress as linear and uniform.

These cities also provoke reflection on humanity's bond with landscape. They were built with awareness of place, responding to rivers, mountains, and climate in ways that sustained life for generations. Their collapses offer lessons about vulnerability, imbalance, and the cost of ignoring environmental limits.

## Echoes That Still Resonate

As the montage fades and modern skylines reappear, the continuity of America's ancient landscapes becomes clear. Highways trace former trade routes. Cities rise atop ceremonial grounds. Rivers still carve the valleys where early civilizations anchored their worlds. Even when structures vanish, patterns endure in geography, memory, and cultural inheritance.

The lost cities of the Americas are not truly lost. Their echoes persist in the stones left behind, the stories still told, the landscapes shaped by their presence, and the knowledge preserved by descendant communities. They reveal that human creativity is as old as the continents themselves, and that every abandoned plaza or buried foundation holds lessons about resilience, innovation, and the fragile balance between people and the environments they shape.

In their silence, these cities speak. Their stories, once hidden beneath forests, canyons, deserts, and time, continue to guide understanding of America's past and illuminate possibilities for the future.

Made in the USA
Monee, IL
04 November 2026

67381310R00095